HERAUSGEBER EDITOR
WOLF D. KARL

COLOURS FOR HOTELS

酒店色彩教程

（德）阿克塞尔·维恩（Axel Venn）
（德）约阿尼纳·维恩·洛斯奇（Janina Venn-Rosky） 著
（德）科琳娜·克雷奇马·乔恩克（Corinna Kretschmar-Joehnk）

深圳市艺力文化发展有限公司 译
深圳市欣欣嵘珺图书有限公司 策划

江苏凤凰科学技术出版社

ALPHABETICAL INDEX 字母索引

406	African	粗犷非洲	374	futuristic	未来主义	120	rejuvenating	焕发新生
222	antique	古色古香	076	garish	炫丽夺目	132	relaxing	轻松自在
200	appealing	新奇有趣	036	generous	大方怡然	180	restrained	婉约矜持
154	appetising	秀色可餐	104	gentle	体贴入微	364	revolutionary	颠覆传统
168	authentic	返璞归真	148	healing	情感治愈	362	richly accented	浓墨重彩
242	bewitching	妖媚娇艳	034	high-class	高档品质	240	romantic	浪漫多情
332	bohemian	放荡不羁	018	high-end	高端大气	206	rural	山水田园
326	bold	胆大妄为	220	historic	历史沉淀	210	rustic	原始乡土
058	bourgeois	中庸务实	198	homely	家常记忆	208	secluded	与世隔绝
418	British	纯正英式	372	iconic	标志意义	138	sensitive	敏感细腻
420	Central European	唯美中欧	238	idyllic	诗情画意	334	sensual	感官享受
388	charming	妩媚多姿	268	imaginative	想入非非	136	serene	静谧安详
344	chic	潇洒别致	270	impressive	惹人注目	184	simple	简约朴素
292	classic	经典传承	276	inspiring	心驰神往	186	sober	素净质朴
346	colourful	五彩缤纷	290	international	国际风范	324	stark	了无修饰
118	comfortable	舒适惬意	336	intoxicating	令人陶醉	308	snug	舒适温暖
064	conservative	保守克制	122	invigorating	万物生长	134	soothing	轻柔舒缓
306	cosy	亲密无间	264	lively	活泼灵动	046	sophisticated	精致得体
266	creative	天马行空	014	luxurious	豪情奢华	404	South American	奔放南美
078	dandified	狂欢浮华	400	Mediterranean	地中海风	350	spectacular	叹为观止
092	discerning	谨慎挑剔	278	motivating	激发斗志	050	splendid	金碧辉煌
392	discreet	低调慎重	288	multi-cultural	多元文化	124	stimulating	振奋人心
170	down-to-earth	朴实无华	166	natural	回归自然	182	strict	一丝不苟
254	dreamy	如梦如幻	110	neat	简洁淡雅	060	stylish	时尚现代
422	East European	斑斓东欧	048	noble	隐性贵族	282	tempting	鲜艳夺目
376	eccentric	离经叛道	322	nonconformist	不入主流	090	top-notch	首屈一指
032	elegant	优雅内敛	416	North European	纯美北欧	196	touristic	爱在旅途
244	enchanted	魔法情缘	310	nostalgic	温婉怀旧	360	turbulent	汹涌澎湃
088	enriching	丰富多彩	402	Oriental	东方情调	348	unusual	与众不同
280	entertaining	娱乐至上	390	plain	平淡原味	252	uplifting	引人向上
338	exaggerated	潮流浮夸	226	plush	豪华舒适	062	upscale	卓越上层
020	exclusive	私享定制	256	poetic	朦胧诗意	164	village-like	乡村风格
094	expensive	贵气逼人	072	premium	超凡优选	152	vitalising	无限活力
366	extravagant	肆意放纵	016	prestigious	尊贵华美	320	weird	捉摸不定
304	fairytale-like	恍如童话	044	princely	皇室隽永	386	witty	妙趣横生
194	family-friendly	居家温情	378	progressive	激进革新	074	world-class	世界一流
030	fascinating	引人入胜	212	puristic	纯粹至美			
250	festive	节日颂歌	108	quiet	静水深流			
224	feudal	盛世王侯	150	refreshing	清凉酷爽			
294	folkloristic	地域风俗	106	regenerative	生机盎然			

CONTENTS

Chapter 1

- 004 Preface / Introduction
- 012 **Hotels with luxury appeal**
- 014 luxurious - prestigious - high-end - exclusive
- 030 fascinating - elegant - high-class - generous
- 044 princely - sophisticated - noble - splendid
- 058 bourgeois - stylish - upscale - conservative
- 072 premium - world-class - garish - dandified
- 088 enriching - top-notch - discerning - expensive

Chapter 2

- 102 **Hotels with regenerative factor**
- 104 gentle - regenerative - quiet - neat
- 118 comfortable - rejuvenating - invigorating - stimulating
- 132 relaxing - soothing - serene - sensitive
- 148 healing - refreshing - vitalising - appetising

Chapter 3

- 162 **Hotels – with a host family**
- 164 village-like - natural - authentic - down-to-earth
- 180 restrained - strict - simple - sober
- 194 family-friendly - touristic - homely - appealing
- 206 rural - secluded - rustic - puristic
- 220 historic - antique - feudal - plush

Chapter 4

- 236 **Hotels between dream and poetry**
- 238 idyllic - romantic - bewitching - enchanted
- 250 festive - uplifting - dreamy - poetic
- 264 lively - creative - imaginative - impressive
- 276 inspiring - motivating - entertaining - tempting
- 288 multi-cultural - international - classic - folkloristic
- 304 fairytale-like - cosy - snug - nostalgic

Chapter 5

- 318 **Hotels for connoisseurs and Bohemians**
- 320 weird - nonconformist - stark - bold
- 332 bohemian - sensual - intoxicating - exaggerated
- 344 chic - colourful - unusual - spectacular
- 360 turbulent - richly accented - revolutionary - extravagant
- 372 iconic - futuristic - eccentric - progressive
- 386 witty - charming - plain - discreet

Chapter 6

- 398 **Hotels of the regions**
- 400 Mediterranean - Oriental - South American - African
- 416 North European - British - Central European - East European
- 430 Index

目录

序言 / 引言

奢华魅力酒店

豪情奢华 – 尊贵华美 – 高端大气 – 私享定制
引人入胜 – 优雅内敛 – 高档品质 – 大方怡然
皇室隽永 – 精致得体 – 隐性贵族 – 金碧辉煌
中庸务实 – 时尚现代 – 卓越上层 – 保守克制
超凡优选 – 世界一流 – 炫丽夺目 – 狂欢浮华
丰富多彩 – 首屈一指 – 谨慎挑剔 – 贵气逼人

生机盎然的酒店

体贴入微 – 生机盎然 – 静水深流 – 简洁淡雅
舒适惬意 – 焕发新生 – 万物生长 – 振奋人心
轻松自在 – 轻柔舒缓 – 静谧安详 – 敏感细腻
情感治愈 – 清凉酷爽 – 无限活力 – 秀色可餐

家庭式酒店

乡村风格 – 回归自然 – 返璞归真 – 朴实无华
婉约矜持 – 一丝不苟 – 简约朴素 – 素净质朴
居家温情 – 爱在旅途 – 家常记忆 – 新奇有趣
山水田园 – 与世隔绝 – 原始乡土 – 纯粹至美
历史沉淀 – 古色古香 – 盛世王侯 – 豪华舒适

梦幻诗意酒店

诗情画意 – 浪漫多情 – 妖媚娇艳 – 魔法情缘
节日颂歌 – 引人向上 – 如梦如幻 – 朦胧诗意
活泼灵动 – 天马行空 – 想入非非 – 惹人注目
心驰神往 – 激发斗志 – 娱乐至上 – 鲜艳夺目
多元文化 – 国际风范 – 经典传承 – 地域风俗
恍如童话 – 亲密无间 – 舒适温暖 – 温婉怀旧

鉴赏家与不羁者的酒店

捉摸不定 – 不入主流 – 了无修饰 – 胆大妄为
放荡不羁 – 感官享受 – 令人陶醉 – 潮流浮夸
潇洒别致 – 五彩缤纷 – 与众不同 – 叹为观止
汹涌澎湃 – 浓墨重彩 – 颠覆传统 – 肆意放纵
标志意义 – 未来主义 – 离经叛道 – 激进革新
妙趣横生 – 妩媚多姿 – 平淡原味 – 低调慎重

地域风情酒店

地中海风 – 东方情调 – 奔放南美 – 粗犷非洲
纯美北欧 – 纯正英式 – 唯美中欧 – 斑斓东欧

索引

PREFACE BY THE EDITOR WOLF D. KARL

Hotels are the oases of rest for a nomadic society which has declared the entire globe to be home. After having become more or less accustomed to sedentarism over the last ten thousand years and having adapted to stable conditions, since the beginning of the third millennium, the world's peoples have entered a phase of a worldwide and relatively well organised migration. Maximum flexibility, curiosity and a love of experimentation characterise the spirit of the age.

Each temporary dwelling space must feature the same qualities as the four walls in which we live. In precisely this way, the hotel forms the recreational centre in which we can relax and chill out. Hotels as homes must have at least the same feel-good qualities as a private home. The philosophy of life in hotels aims at creating an atmosphere in which friendships are made, an atmosphere which appeals to the senses and which prompts a mood of enjoyment.

In addition to "dolce-far-niente", the pleasures of sweet idleness, the interior processes must also be influenced by invigorating characteristics. The design message always results in two different appeals: it aims at both calming and stimulating. The philanthropic design model combines the emotional with the useful. Sensuous satisfaction is also an expression of physical wellbeing.

Gentle colour shades rather than high-contrast ones, round rather than angular shapes, comfortable rather than stylish design, harmony rather than pure aesthetics, gentleness rather than cool elegance serve us as emotional reflectors for improving the quality of the atmosphere.

序言

编者 Wolf D. Karl

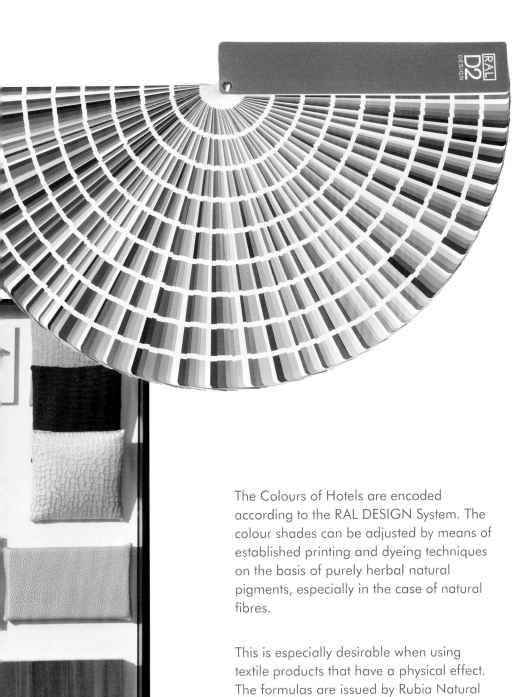

游牧时代的人们总是四海为家,酒店则是人们休憩的绿洲。在过去的成千上万年里,人类逐渐习惯定居,适应了稳定的生活。随着新千年的到来,世界各民族进入到相对有序的全球性迁徙阶段,极大的灵活性、强烈的好奇心和对尝试新鲜事物的热衷,都是这个时代的特征。

临时居所应当具备日常居住空间所具有的品质,就此而言,酒店——供我们放松心情的休闲中心,也应像家一样,让人感到舒心。酒店的理念就是营造一种友爱和调动人们的感官去体验的氛围,让生活在其中的人们身心愉悦。

酒店空间设计不仅要体现"闲适"之趣,还应充满活力。在这种设计语言下,往往会形成两种效果完全不同的空间,要么显得平静,要么显得活泼。具有参考价值的设计不仅实用,更能引起情感上的共鸣,因为感官的满足也是身体健康的表现之一。

色彩柔和、外形圆润、舒适和谐而又温馨的空间,与色彩对比鲜明、外形棱角分明、设计时尚美观而又优雅的空间相比,更能引起人的共鸣,从而提升空间氛围。

书中酒店的色彩是依据RAL色系编码,色彩的深浅可以通过基于纯天然植物颜料的印染技术进行调整,尤其是在天然纤维的情况下。

若这些染料能在纺织布料上起物理反应,效果会更好。荷兰的Rubia Natural Colours曾发表针对纺织产品、皮革、纸张和化妆品的色彩物理反应准则。

The Colours of Hotels are encoded according to the RAL DESIGN System. The colour shades can be adjusted by means of established printing and dyeing techniques on the basis of purely herbal natural pigments, especially in the case of natural fibres.

This is especially desirable when using textile products that have a physical effect. The formulas are issued by Rubia Natural Colours, NL for textile products, leather, paper, and cosmetics.

INTRODUCTION

This book is a design research paper and an academic study, a creative workbook and a set of instructions for inviting, atmospheric and exhilarating ambiences. It is also an analytical fact book and, equally, an experimental design template. The colour-analytical part of this book is based on an academic study that focuses on colour combinations in hotels already designed or planned, as well as on further subject-related design contents. For this, we analysed 120 terms from the hotel world according to colour, shape and content profiles. We wanted to know in advance exactly what, for example, "international", "nostalgic", "luxurious", "discreet", and "plain" look like. On page after page, we develop recipes for design; the research paper functions like a recipe book and can be used to stimulate, refine, reduce and perfect.

Our main objective is to create a hotel book with beautiful rooms in which one simply feels comfortable, whether it is 5-star accommodation or a homely country hotel with all imaginable amenities, whether it is a 7-star hotel de luxe for minor and major millionaires, or whether it is simply a rustic mountain chalet or a stylish small town house with a home-like setting. We want to show a broad range and present both simple and fascinating interiors.

By way of examples we want to find out why we might prefer staying in a hotel with a classic style to staying in what is explicitly a design hotel. We are looking for the minor and major competitive differences. The reason why we prefer staying in this or that hotel can rarely be reduced to the question of it being cheap or expensive. A warm welcome does not equate to a room at the right temperature. An excellent chef is no guarantee for comfortable seating in the restaurant. Sparkling clean and neat guest rooms are not actually evidence of friendly service. It is the sum of its various characteristics that makes staying in a hotel an unforgettably pleasant experience or a nightmare.

The classification system separating hotels into categories may serve as an indication of satisfaction or lack of satisfaction, of the actual quality of one's stay and of its facilities. However, what does the best internet equipment or the most splendid hammam, the most elegant fitness centre and the freestanding bathtub in the middle of the hotel room mean to us if we are neither internet junkies, nor wellness-oasis fans, fitness freaks or living-room-slash-bathroom enthusiasts? We do know that even the hotel guest who is critical of alcoholic drinks will not avoid the cosy hotel bar if its design is atmospheric and if it serves as a meeting place for the elegant, the beautiful, the interesting and the charming, inviting all to linger, talk and listen to music.

The most pleasant hotels are the ones that combine communication and distance, aesthetics and dignity, opulence and reduction, zeitgeist and nostalgia, calm and stimulation, urbanity and homeliness, cordiality and functionality, individual service and collective attention.

The most important quality and feel-good characteristics that constitute a hotel are references to the location and the landscape, to architecture, interior design, in-house culture and service. Numerous other elements certainly also rank among the essential attributes that turn a hotel into a symbol; among these, the emotional contents are always at the top of the satisfaction scale.

Human beings are at the centre of our academic and design-oriented study. We take them very seriously, and that is why we take them under our philanthropic wing. Their idiosyncrasies and peculiarities make us grow even fonder of them.

引言

本书既可称作设计研究的学术论文,又可充当创意手册和空间设计指南。书中不仅分析了大量实例,而且提供了试验性的设计范本。依据对现有酒店色彩搭配以及其他相关设计领域的学术研究,书中用大量的篇幅作了色彩分析。还从色彩、外观和内容等角度分析了116个形容酒店的短语。我们都想提前知道"国际风范"、"温婉怀旧"、"豪情奢华"、"低调慎重"和"平淡原味"等色彩主题究竟各有怎样的风采。本书与其说是一份研究报告,不如说是一套色彩搭配教程,因为书中每一页都为设计提供了配色方案。阅读此书能启发灵感,升华设计,更能化繁为简,完善设计。

我们旨在以舒适、美观的酒店空间为案例,创作一本酒店类的书,从五星级酒店到设施齐全的乡村酒店,从专为富翁打造的奢华七星级酒店到朴素的山间农舍,抑或是风格别致的家庭式小镇酒店,书中应有尽有。

我们为何更愿住古典风格的酒店,而不愿住设计感强的酒店?书中的案例将给予我们答案。酒店之间的竞争力为何存在差异?这也是我们试图解答的问题。酒店的选择并不完全取决于价格的高低;热情的态度并不意味着客房的温度适宜;出色的厨师并不能保证餐厅内的座椅舒适;亮堂整洁的客房也不等同于友善的服务。书中总结了各种空间的特点,告诉你哪些酒店会令你印象深刻,哪些酒店会给你噩梦般的感觉。

酒店分级系统可以反映顾客的满意程度、顾客此次体验的实际质量以及酒店设施的质量。然而,如果我们既不是网虫,也不是保健粉,更不是健身狂和泡澡狂,那么最好的互联网设备、最华丽的蒸汽浴室,甚至是最雅致的健身中心、最豪华的独立浴缸,对我们来说真的那么重要吗?但我们知道,即使是禁酒论支持者,也无法拒绝酒店里氛围良好、舒适惬意的小酒吧。因为它被设计得如此迷人优雅,充满乐趣,吸引着人们到此聚会、交谈、聆听音乐,甚至使人们流连忘返。

备受青睐的酒店不仅知晓如何亲近客户,了解其需求,而且懂得把握与客人之间的距离。这些酒店的设计美中略显尊贵,丰富而简练,时尚而怀旧,沉稳而活泼,文雅而质朴,温馨而实用,不仅提供个性化服务,同时提供集体关怀。

对于酒店来说,地域特色、景观风情、建筑外观、室内设计、内部文化和服务等方面都是最重要的品质,只有这些方面都考虑到位,这家酒店才会成为人们的首选,甚至成为城市标志。当然,还有许多其他重要元素,其中与客人满意度联系最紧密的莫过于酒店的人文关怀。

Preface / Introduction

Only rarely do we like, accept and love our friends because their intelligence and reason appeals to us. Instead, we are enchanted by their droll otherness, creativity, their imaginativeness as storytellers and their originality.

In this book, we disclose many a secret wish. We learn more from people's feelings than from their cognitive and reality-related responses. Our research and studies, their decoding, the expertise we have acquired, the love of detail and of the grand design are the subject of this book.

We hope that the readers of this book will make keen use of its creative contents. The enclosed template helps readers in that it makes it easier to focus on idividual colour fields. We have attached great importance to illustration and product details in order to demonstrate that it is indeed the case that individual results are preconditions for the success of overall impressions.

The Dutch philosopher Huizinga suspected that play is the foundation of culture. We believe that attempts, experimentation and even chaotic trials are just as essential to masterly achievements as is the intellectual approach.

在我们此次以学术和设计为导向的研究中，人是中心。我们十分关注人，将其视为服务对象。而人的种种个性和特点更能激起我们的好奇心，从而促使我们投入更多的关注。

我们往往不会因某人的才智和理性，而是因为他的幽默感和创造力，以及说书人般的想象力和独创性，与其结为好友。

本书暗含了许多期许，相对人们的认知和现实反应而言，人们的情感往往能透露更多的信息。这次调查和研究的结果与解答的问题，以及我们能从中获得的专业知识，还有对细节的挑剔以及对整体设计的掌控，都是本书的主题。

我们希望读者能充分利用书中富有创意的内容。书内提及的方案能够帮助读者更加轻松地聚焦独立色彩应用领域。我们侧重插画和产品细节展示，以此强调细节效果的成功才是整体空间表现获得成功的先决条件。

荷兰哲学家Huizinga曾怀疑，奠定文化的基础不是其他，而是游戏。我们相信，努力去做、去试验，甚至胡乱地尝试，这些充满智慧的过程，都是促成伟大成就的关键。

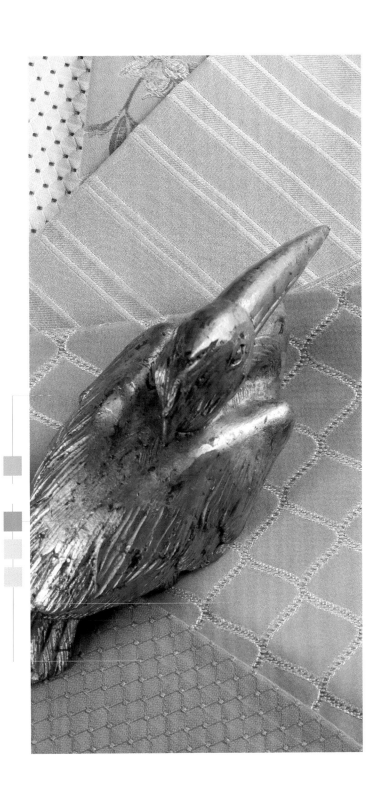

Preface / Introduction

THE TEMPLATE

配色板

The combined creative fields of theme groups list the essential colours of each individual theme. Highly application-oriented colour schemes can be developed with the help of the template. This works in a playful, experimental way for all colour-related terms represented in this book.

The template has a black and a white side with good reason. Very bright shades can be better determined using the white template: it modulates the single light shades in a more nuanced manner. The black template, on the other hand, brings medium to dark colours more densely and more harmoniously together. It is not suited for very bright colours because the contrast of the extremely strong bright and the dark blurs the nuances.

The template also makes possible the creation of colours that span a number of themes. Colour groups from the reproductions of the research study can also be "searched". Frequently, this brings about surprising, creative and useful results.

每个主题框架下都有一个组合而成的创意版块，这里列出了每个主题相对应的关键色彩。按照这个版块，就可制订出可实施性较高的用色方案。此外，本书还通过有趣的实验方式来阐释书中提到的所有色彩描述词。

配色板以合理的逻辑安排出黑色系区域和白色系区域。白色系区域能细致入微地对单种浅色调进行调制，从而能够更好地界定明亮的色彩；而黑色系区域能将中度色调至深色调变得更浓厚、更协调。这个区域不适合特别明亮的色彩，因为这种极强的明亮度与暗色调放在一起时，产生的对比会模糊掉这些色彩的细微差别。

此外，在配色板上也能跨主题配色。本书通过调查研究得出，颜色主题依然可以进行"研究"。这个过程常常带来更多的惊喜、创意和实用的成果。

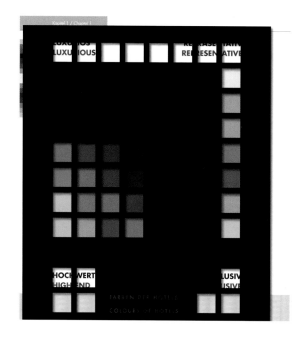 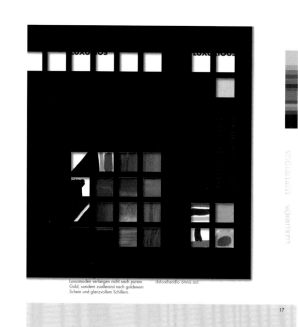

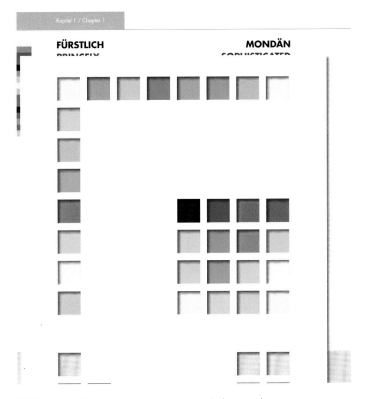

White template colour range at right angles
白色系区域分布在右上角区域

Template on the painted fields: search for harmony
配色板中涂色的区域用于寻求和谐

Black or white template in the same position: a completely different picture appears.
同一色块中的白色系或黑色系区域形成完全不同的画面。

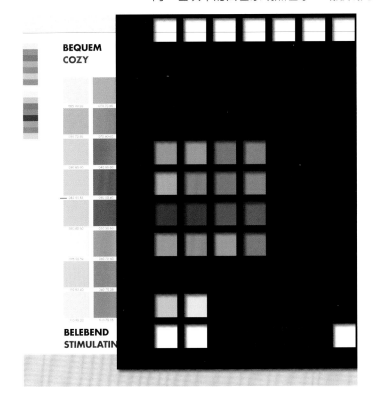
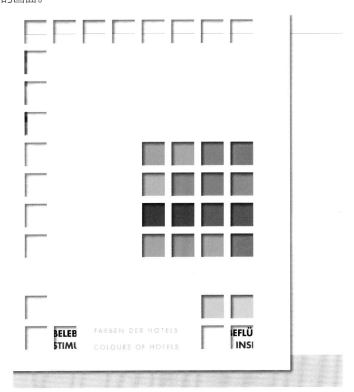

/ Chapter 1

HOTELS WITH LUXURY APPEAL: FROM THE LUXURIOUS TO THE EXPENSIVE

There is nothing better than being a member of a democratic civil society and living in a castle, if only sporadically. Plenty of stucco on the ceilings, rooms with ceilings four and a half metres high, a freestanding bathtub with lion-paw feet, brocade in front of the three-metre high windows, tassels and cord edges on cushions and upholstery, as well as baroque, rococo or Empire armchairs that offer ample room for two people. Bordeaux-red fabric covers the walls. On the sideboard stands a huge vase with fresh gladiolas.

At the table d'hôte, the guests come together in the evening in sober suits and elegant dresses; trouser suits are frowned upon, as are casual athlete's uniforms with logos on their collars. Even if the great meal includes 14 to 20 people of various age groups and different nationalities, the evening will, thanks to skilful moderation by the host or hostess, be judged a success due to entertaining conversations and charming chats.

Of course the menu, the wines, the table decoration, the ambience and the gentle lighting, supported by lighted candles, should also contribute to making the evening unforgettable.

Small private hotels in France, England and Italy still cultivate this form of hospitality. Why do they have to remain an insider's tip? Imitation is definitely in order. Quality of living is always a question of extraordinary ambience, of exceptional guests and a dedicated hostess, but also of beautiful colours and colourful anecdotes.

Such small and intimate hotels also reveal themselves to be guarantors of informal understanding among nations because the guests are invited to participate in multi-national dinners. It is amusing when 16 people sit at a table with an joint verbal treasure of two to three dozen living languages, and have great conversations. There is always an interpreter translating from one language into another while the next one conveys another idiom. Impatience would not be appropriate, and neither would be the certainty that one can exclude the possibility of a misunderstanding.

奢华魅力酒店：
从豪情奢华到贵气逼人

作为民主社会中的一员，如果能偶尔入住城堡，那就最好不过了。想象一下，住在高达4.5米的房间里，天花板上有着厚厚的灰泥，房间内设有狮爪状腿的独立浴缸，3米高的窗户前挂着锦缎窗帘，还有精致的流苏、坐垫和抱枕，以及巴洛克式、洛可可式或是帝国时代风格的豪华双人扶手椅，餐具柜上摆放着插有新鲜剑兰的大花瓶，甚至墙壁也为枣红色织物包裹。

夜幕降临时，身着笔挺西装和高雅礼服的男女宾客来到宴厅，身着长裤套装的女宾则会引起人们的不满，因为那些衣领处贴着标签的长裤套装被认为是休闲运动服装。宴会上，如果主持人能娴熟地调动气氛，妙语连珠，即使这次宴会只有14到20位宾客，即使宾客的年龄和国籍各不相同，这也会被认为是一次成功的宴会。

当然，这个夜晚之所以如此难忘，也需归功于菜肴、美酒、餐桌装饰和宴会氛围，以及柔和的灯光、曼妙的烛光。

至今，英、法、意的私人小酒店仍很注重这种待客形式。那么，为何这种形式只对内公开？自然是为了杜绝模仿。酒店品质往往与独特氛围、特殊顾客以及专属服务员等有关，当然也不能忽略出色的色彩搭配和精彩的趣闻轶事。

在这些亲密度高的小酒店里，客人常常会被邀请到跨国宴会中，因此它们还能以这种非正式的方式促进国家之间的相互理解，当持两到三种不同语言的16位宾客齐聚一堂、愉快交谈时，总得有人把一种语言转换成另一种语言，并且阐释习语，这是非常有趣的场景。这也要求人们非常耐心，以避免引起误会。

014	luxurious - prestigious - high-end - exclusive	豪情奢华 – 尊贵华美 – 高端大气 – 私享定制
030	fascinating - elegant - high-class - generous	引人入胜 – 优雅内敛 – 高档品质 – 大方怡然
044	princely - sophisticated - noble - splendid	皇室隽永 – 精致得体 – 隐性贵族 – 金碧辉煌
058	bourgeois - stylish - upscale - conservative	中庸务实 – 时尚现代 – 卓越上层 – 保守克制
072	premium - world-class - garish - dandified	超凡优选 – 世界一流 – 炫丽夺目 – 狂欢浮华
088	enriching - top-notch - discerning - expensive	丰富多彩 – 首屈一指 – 谨慎挑剔 – 贵气逼人

/ Chapter 1

LUXURIOUS

豪情奢华

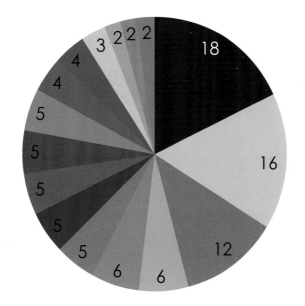

Luxury is not necessary. Its non-existent affinity to real user value in particular makes it so desirable. The large black section acts as a stable statement of its power.

奢华在人们的生活中并非必需品,因为在用户的价值观中,它并没有亲和力,然而正是如此人们才更向往奢华。图中大块黑色区域强有力地标明黑色在这一色彩搭配中的影响力。

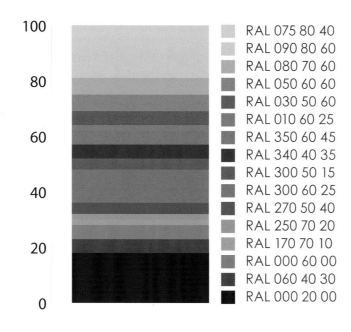

This is what colour combinations of outrageously expensive silk scarves often look like. One wears them as a trophy and as a conspicuous fluttering signal of recently acquired or recovered authority.

这种色彩搭配往往会用在极为昂贵的丝巾上。人们会戴这种丝巾来显示身份,或者把它当作一种显性信号,来说明自己最近获得或重获了某种权力。

The drawing is not as necessary as the colours: a few rough lines and simple basic shapes from round to angular suffice.

左图中的色彩比图形更具表现力:粗糙的线条勾勒出圆形和菱形等简单而基本的形状。

LUXURIOUS 豪情奢华

Gold, black, violet and pink and rose are pure luxury. Luxury depends on glamour and not necessarily on tangible splendour. Luxurious fashion does not call for pure gold but first and foremost for golden looks and glamorous opalescence.

金色、黑色、紫色、粉红色和玫瑰色都是奢华色调。奢华并不在于表面的富丽堂皇，而在于潜在的魅力。奢华的时尚并不要求使用足金，它至少应是金色并泛着蛋白色的光泽。

/ Chapter 1

PRESTIGIOUS

尊贵华美

Metallic colours like brass, copper and gold as well as the primary colours red, blue and yellow and their derivatives have long been among to the preferred hotel colours all over the world.

黄铜色、红铜色和金色等金属色系，红、蓝、绿三原色及其派生色彩，是世界各地的酒店惯用的色彩。

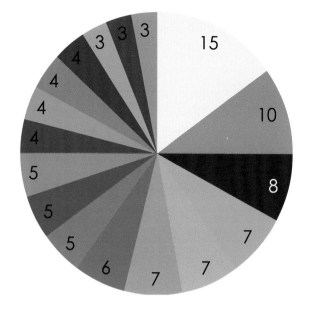

The colour range is powerful and radiates much of the authority required when giving directives. If a reception situation is designed with this colour scale, the guests will feel as if they are with Gloria Swanson or Prince Charles having afternoon tea.

这一色域的色彩极具表现力，当用于下达指令时，能够显示出必要的权威性。如果一场接待活动以这一色域来设计，来宾将感受到与诸如葛洛丽亚·斯旺森或查尔斯王子等尊贵人士同享下午茶的荣耀感。

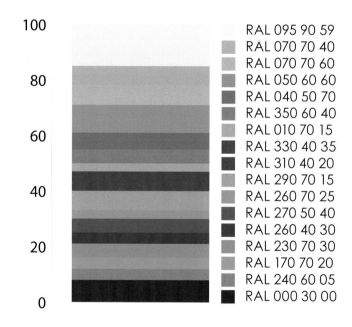

RAL 095 90 59
RAL 070 70 40
RAL 070 70 60
RAL 050 60 60
RAL 040 50 70
RAL 350 60 40
RAL 010 70 15
RAL 330 40 35
RAL 310 40 20
RAL 290 70 15
RAL 260 70 25
RAL 270 50 40
RAL 260 40 30
RAL 230 70 30
RAL 170 70 20
RAL 240 60 05
RAL 000 30 00

The signals appear demonstrative and without flourishes, dignified and exalted. Cordiality or special sympathies are not visible.

这种色彩发出的信号明确，庄重而高贵，且不会太花哨，当然也不带一丝情感。

PRESTIGIOUS　　　　尊贵华美

The colour programme dispenses with surprises: it is suitable for the masses and thus tolerable and comprehensible for everybody. All shades that might possibly cause irritation, like brown, beige and yellow-green, are absent. The "in-between" colours lack an unambiguous ability to express value.

　　该色彩方案配色沉稳，往往不会有太大惊喜。它易于让人们接受和理解，适用于面向大众的设计项目。所有可能引起人们不悦的色彩，如棕色、米黄色和黄绿色等，不适用于该方案。这些中间色由于自身色彩界定不明，可能无法完全传达设计本质。

/ Chapter 1

HIGH-END

高端大气

Rooms designed in such colour schemes feature a high capacity for luxury. These shades call for perfect materials and products and value-preserving careful craftsmanship.

采用这种配色方案的空间会显得十分豪华，也只有完美的材料、精美的物品以及精湛的工艺才能与这些色彩相衬。

The discerning colour range is more than a statement. It is a challenge to the designer to make the good even better.

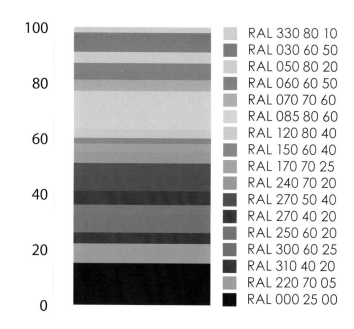

RAL 330 80 10
RAL 030 60 50
RAL 050 80 20
RAL 060 60 50
RAL 070 70 60
RAL 085 80 60
RAL 120 80 40
RAL 150 60 40
RAL 170 70 25
RAL 240 70 20
RAL 270 50 40
RAL 270 40 20
RAL 250 60 20
RAL 300 60 25
RAL 310 40 20
RAL 220 70 05
RAL 000 25 00

该色域中的色彩对比鲜明，很具表现力。对设计师来讲，要把这些色彩应用到炉火纯青的境界，极具难度。

The signs, too, are expressions of stability and power. The horizontal stripes in particular announce permanence and security.

同样，这些图形显得非常稳定，极富力量感；尤其是横纹，体现出一种永久性和安全感。

HIGH-END 高端大气

The dark colours and golden shades have always been more expensive than humble beige or ailing light green. Deep black, too, is always a symbol of power, strength and aristocratic distance.

深色系和金色系往往会比浅米色、亮绿色和深黑色显得更贵气，虽然深黑色通常都是权力、力量和贵族距离感的象征。

EXCLUSIVE

私享定制

The shades vie with each other in a power struggle. The contrariness and ambiguity are glaringly obvious. Attraction and rejection as a design model! Soot-black versus light yellow, orange-red and purple.

此色系中的各种色彩争奇斗艳、不相上下，而且鲜明的对比和不明确性兼具。以这些色彩设计出的空间兼具吸引力和排斥感，比如炭黑色搭配浅黄色，橙红色搭配紫色。

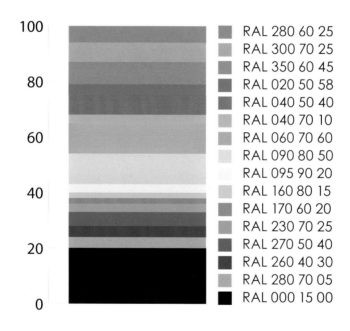

The colour range is ambitious and fashionable, but also traditional, baroque and trendy. The outrageously expensive, funfair punk, new swank, and old dignity are united.

此色域有的色彩鲜亮夺目，时尚又略显传统，华贵又略显俏皮，新风尚中也融入了传统的庄严感，同时还透露出一丝巴洛克风范。

Vibrant, graphic signs stand for exclusivity. We are not surprised to recognise them.

我们会发现这些活力四射的图案代表着"独一无二"。

EXCLUSIVE   私享定制

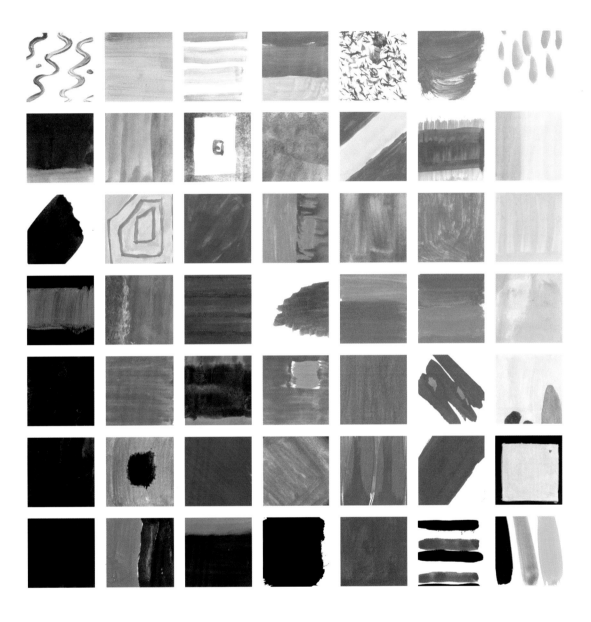

The exclusive imposes a ban on what we describe as "not belonging". Being banned from participation is especially hard for those who insist on exclusiveness. The colour substance of this term ranges somewhere between Bollywood and Hollywood.

该色域中的色彩强烈反对"非专属色"的加入。坚持要独有而被禁入,是极其残酷的。此色域中的色彩跨度较大,就像宝莱坞和好莱坞一样,其间相差甚远。

/ Chapter 1

LUXURIOUS
豪情奢华

PRESTIGIOUS
尊贵华美

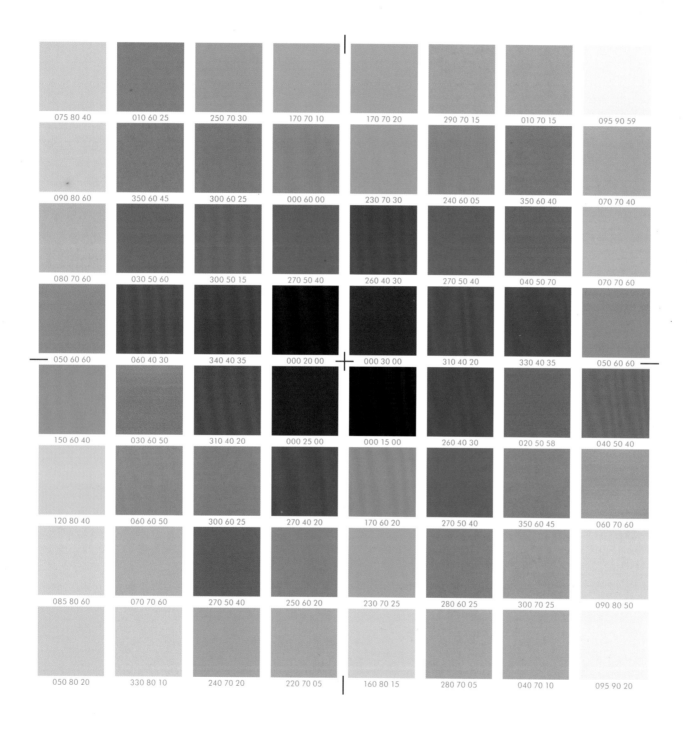

HIGH-END
高端大气

EXCLUSIVE
私享定制

LUXURIOUS
PRESTIGIOUS
HIGH-END
EXCLUSIVE

豪情奢华
尊贵华美
高端大气
私享定制

Doing without luxury can be as excruciating as an overabundance of pink, splendour, pomp, purple, rosewood, powder, chocolates and plush. Looking closely at the adjacent 64-shade colour field, it becomes clear that the dark shades together with the bright, pure pastel hues form an equal community of feminine and masculine nuances. Sweet fruit and light blossom shades constitute a unit of splendour, elegance and captivating freshness, whereas the contrary colours are made up of martial black nuances, sophisticated purple and velvet shades, as well as regal shades of blue and gemstone colours.

And yet behind all colours lies an overt claim to power and importance. The Roman emperors wore purple, the French fleur-de-lis stands out resplendently against a blue background, the Russian tsars loved green, and the Dutch court banks on orange because the Orange Order had a lot to do with Oranje. At the Spanish court under Philip II, there was only one shade: majestic black.

过分奢华的空间，与没有丝毫奢华感的空间一样，都令人难以接受，这主要体现在粉红色、紫色、红木色、粉末白和巧克力色等色彩的堆积，以及过分富丽堂皇的装饰上。仔细观察64色卡中色彩的过渡位置，我们不难发现，深色系与柔和的纯亮色系搭配会同时散发出阴柔气息和阳刚气息，显得十分和谐。此外，甜果色和浅花色搭配不仅辉煌典雅，而且清新迷人。而对立的色彩不宜重复使用，如深浅不同的武术黑、色彩相近的紫色和天鹅绒色，以及突显贵气的蓝色和宝石色。

然而，所有的色彩都曾被公然宣称是某种势力和某种价值的代表，例如：罗马帝王钟爱紫色；法国王室的鸢尾花纹章在蓝色的背景下显得更为耀眼；俄国沙皇偏爱绿色；荷兰皇庭以橙色作为标志，因为"橙带党"与奥兰治密切相关；腓力二世统治下的西班牙皇庭则仅以威严的黑色作为代表色。

LUXURIOUS 豪情奢华

050 60 60   075 80 40   000 60 00   000 20 00

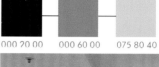
000 20 00   000 60 00   075 80 40

A lobby enveloped by black matt and glossy effects: those who take a seat here will for a moment retreat into a world that evokes a feeling of dignity and significance. This is also illustrated by the wall products at the bottom of this page: subdued metallised effects with a reduced textured surface emphasise man's intentionally reserved attitude towards the product.

Each colouring, like each formal content, black back-stitched-leather sofas deliver a message that seems on the surface to be neutral. In actual fact, it can have an exciting, provocative and rather personal side to it. At some point, the red colour theme must appear, even if it shows only as the piping of a panel of fabric or narrow edge binding. The next colours that suit luxury are yellow, gold and silver.

300 50 15   075 80 40   000 60 00

大堂用黑色亚光和亮光表面来搭配，显得十分威严和高贵。本页页底端图中的亚光金属色泽和布纹外观壁饰品，体现出一种刻意保守的态度。

每种色彩都有自身的魅力，就像每件正式物件一样。黑色倒缝皮革沙发外观看似中性，但实际上却个性鲜明，十分醒目。在有些地方红色主题必须出现，即使是布料上的小块红色包边或镶边。接下来最能体现奢华风范的色彩要数黄色、金色和银色了。

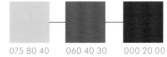

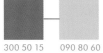

300 50 15   090 80 60

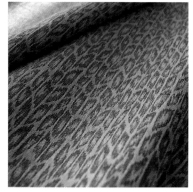

Frequently, too much light is unsuitable. Concentrating on a few areas is more fitting. That way one not only flatters the person but also the colours and materials in the room. It is not without good reason that bars are so sparsely lit: this transforms the modest, hardworking city dweller into a shadow-casting Spanish hidalgo, and the friendly woman next door into a femme fatale.

One has no choice but to love luxury hotels. This is where we find a space that gives us the freedom to occasionally make a wonderful dream of the exceptional come true. Being surrounded by more space than is actually necessary is part of the wonderful world of abundance.

300 50 15   075 80 40

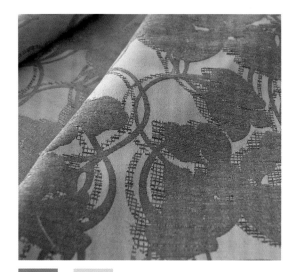

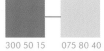

300 50 15   075 80 40

000 60 00   030 50 60   000 20 00

通常，空间不宜过度照明，而应突出少数几片区域，这样不仅能愉悦人心，而且能突出空间的配色和选材。酒吧空间灯盏稀疏，这也并非没有理由，因为这样能让谦逊有礼、勤奋刻苦的城市居民变成颇有影响力的西班牙绅士，让友善的女邻居变成"蛇蝎美人"。

人们都会情不自禁地爱上奢华酒店，因为这个地方能给予我们自由，让我们的美梦变为现实。精彩世界的丰富性还体现在拥有超出实际需求的空间。

075 80 40

000 20 00

080 70 60

050 60 60

300 60 25

075 80 40

030 50 60

000 20 00

000 60 00

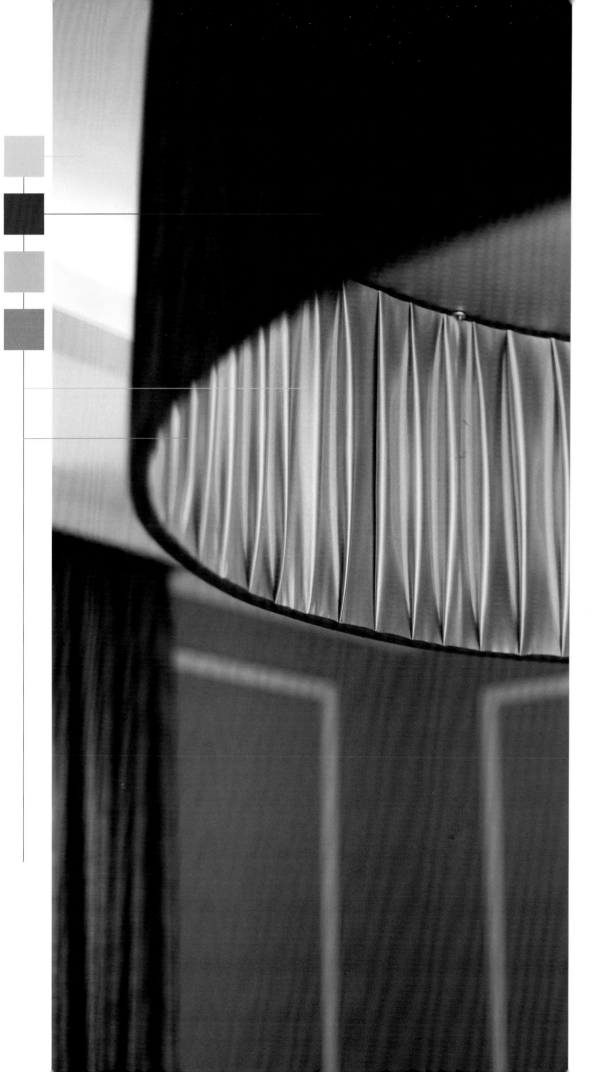

000 20 00

075 80 40

000 60 00

The play with light also has a great impact on the generously designed ambience. Many concentrated but gently flowing lights constitute visual events whose enticing offerings invite you to linger. Warm light attracts people, whereas glaring light arouses their curiosity or repels them.

灯光设计对整体氛围的营造十分关键。光线集中且能缓缓流动的灯光往往能带来视觉冲击，吸引人们驻足。暖色灯光能吸引人的目光，而闪耀的灯光则能激起人们的好奇心，或者让人望而却步。

/ Chapter 1

FASCINATING

引人入胜

Mysterious but also exhausted to morbid interpretations seem to have a sufficient level of interpretational sovereignty. The pompous or extremely precious is mostly not suspected behind the fascinating.

这些充满神秘感的色彩似乎拥有十足的主导权，而不需要过多的无力阐释。毋庸置疑，这些迷人的色彩尽显华贵。

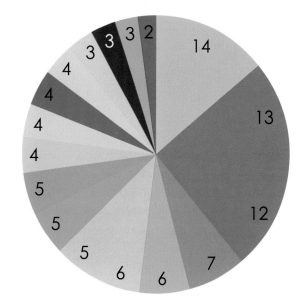

The chromatic colour sequence can be described as "elegant" and "fashionable", "extremely decorative" and "becoming". Sobriety and tradition are relevant attendant aspects which seem to take the place of pomp and showiness.

图中彩色系列显得优雅且时尚，装饰效果好，搭配能力强。然而，虽说这些色彩华丽耀眼，但流露出来的更多的是古朴与清朗。

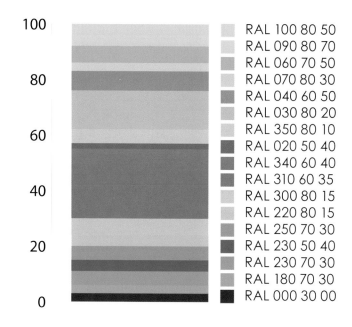

RAL 100 80 50
RAL 090 80 70
RAL 060 70 50
RAL 070 80 30
RAL 040 60 50
RAL 030 80 20
RAL 350 80 10
RAL 020 50 40
RAL 340 60 40
RAL 310 60 35
RAL 300 80 15
RAL 220 80 15
RAL 250 70 30
RAL 230 50 40
RAL 230 70 30
RAL 180 70 30
RAL 000 30 00

It is the gentle chaos and is not the quiet and balance that we find fascinating as a contrast to the consistently straight and compact picture.

"引人入胜"系列色的代表图稍显混乱，与显得平静而匀称的封闭直线图案刚好相反。

FASCINATING 引人入胜

Is our idea of fascination actually not as impressive and magnificent as we should like to believe or anticipate? Maybe a tired metaphor lies behind the concept of the fascinating, which no longer exists in a world of new superlatives that appear on a daily basis.

以上阐释是否颠覆了我们对"引人入胜"的通常的理解？觉得它并没有预想的那么华丽，那么引人注目，或许"引人入胜"的概念中所隐藏的陈旧的象征意义，在日常生活常见的新的表达中早已不存在了。

/ Chapter 1

ELEGANT

优雅内敛

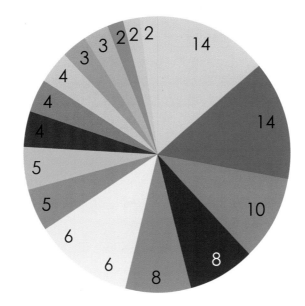

The number of dark shades exceeds those which lie in the middle. The beautiful appearance of radiant glamour, thus the result, is the opposite of elegant distance.

色盘中暗色所占比例超过了中性色彩，这说明优雅色系并不是指那些光亮夺目的色彩。

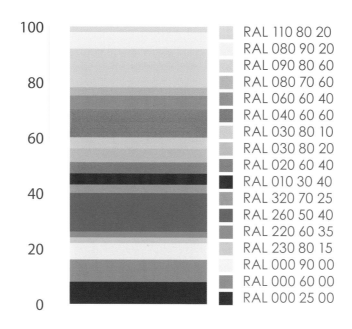

A colour sequence that seems to lack significance. Many shades, neither major nor minor. More wool than silk, more velvet than brocade worked with gold.

RAL 110 80 20
RAL 080 90 20
RAL 090 80 60
RAL 080 70 60
RAL 060 60 40
RAL 040 60 60
RAL 030 80 10
RAL 030 80 20
RAL 020 60 40
RAL 010 30 40
RAL 320 70 25
RAL 260 50 40
RAL 220 60 35
RAL 230 80 15
RAL 000 90 00
RAL 000 60 00
RAL 000 25 00

色卡中的每种色彩都不突出，而且重要性都相当。但关键在于布料的选择，因为与丝绸锦缎相比，羊毛布料和天鹅绒与金色更配，更能体现出优雅的风格。

If anything, the signs are pointing towards a blocky signal quality. Many plain colours, hardly any polychrome interpretations.

或许四个图案之间的共同点就是呈块状。要表达优雅应多用素色，而少用彩色。

ELEGANT   优雅内敛

Nobility likes to hide itself. Here this means that, nuances are of a somewhat ponderous serenity. The abandonment of everything pompous or too loudly coloured is remarkable but adequate and accurate.

高贵人士喜欢隐藏自己，此处意味着"优雅内敛"风格的色彩有些许细微差别，虽显沉闷，仍不失宁静。弃用太耀眼的色彩和太华丽的饰品，可以让空间引人注目却恰到好处。

/ Chapter 1

HIGH-CLASS

高档品质

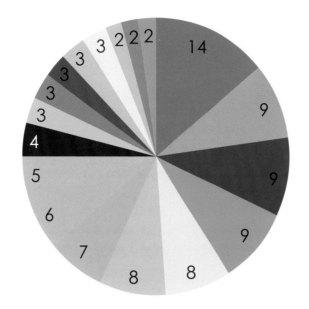

The colour range is dominated by a wide range of shades of blue. Only once in a while do faded, dirty, gold-cognac-green shades appear, the intrinsic value of which does not seem to be significant.

色盘中不同深度的蓝色所占比例最大。虽然也夹杂着一些浓淡不同的金色、棕红色和绿色，但是这些色彩所占比例较小，并不突出。

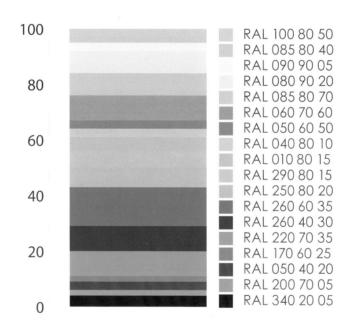

The unaccented colours determine the appearance of the sequence. We perceive it as pleasant, soundless and slightly unsaid to vapid. These are all shades that do not crowd in on us.

色卡中这些并不抢眼的色彩有序地排列着，给人的整体感觉就是舒畅安适、寂然无声，略显枯燥无味，却又不会令人感到压抑。

The pattern is also calm. It does without confusion. Of course, it documents and sublimes virtues of reduction and simplicity.

这些图案显得泰定而不凌乱，体现了简洁直观之美。

HIGH-CLASS　　　　高档品质

With blue clothes, blue cars, blue design people demonstrate a distance towards their own individualised incisiveness. Blue is the metaphor of logic, objectivity and functionality.

身着蓝色服饰，开着蓝色轿车，偏爱蓝色设计的人似乎有意遮掩了自身敏锐的个性。蓝色暗指有条不紊、客观公正、理性务实。

/ Chapter 1

GENEROUS

大方怡然

The colour range has a graphical rather than a picturesque character. In reality it is unspectacular. A little precious and theatrical, without intrusiveness or ingratiation.

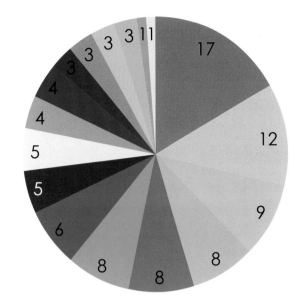

该色盘与其说是一幅多彩的风景画，不如说是一张客观的说明图。实际上，这些色彩并不惹人注目，更无争奇斗艳、曲迎奉承之意，更多的是一丝贵气和盛气。

The result is a light-pastel to lime-coloured range with only few vibrant accentuating shades. It is characterised by aesthetic, seemingly logical substance.

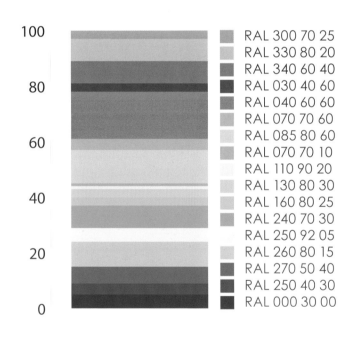

RAL 300 70 25
RAL 330 80 20
RAL 340 60 40
RAL 030 40 60
RAL 040 60 60
RAL 070 70 60
RAL 085 80 60
RAL 070 70 10
RAL 110 90 20
RAL 130 80 30
RAL 160 80 25
RAL 240 70 30
RAL 250 92 05
RAL 260 80 15
RAL 270 50 40
RAL 250 40 30
RAL 000 30 00

色彩由淡色至灰色依次排列，中间夹杂些许鲜亮色彩，层层递进，充满美感。

Straightforward signals are set: blurred diagonal, circular and dot symbols, and every now and then swirling amorphous shapes.

模糊的对角线、环形和点，以及打旋的无定形的图案，都是很直观的标志图。

GENEROUS 大方怡然

Pure shades prevail. Frequently, they are presented to us as airy and cloudy. No brown, hardly any grey or black disturb the clear, partly transparent design.

当设计以纯色为主导时，会给人一种如坠烟海的感觉，棕色无迹可寻，灰色或黑色较为少见，空间会显得明净、通透。

/ Chapter 1

FASCINATING
引人入胜

ELEGANT
优雅内敛

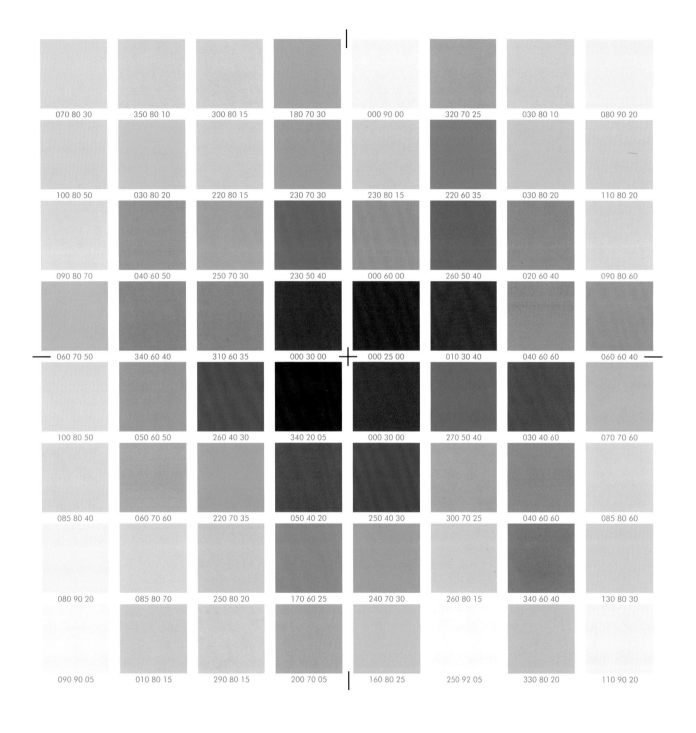

HIGH-CLASS
高档品质

GENEROUS
大方怡然

FASCINATING
ELEGANT
HIGH-CLASS
GENEROUS

引人入胜
优雅内敛
高档品质
大方怡然

The colour range is not necessarily plainer compared to the previous group of four, but its overall appearance is lighter. However, it lacks the powerful pithiness and the renunciation of too much awareness of power. The huge field of blue and green nuances and of rare shades of red points to a subdued colour structure and emotionality.

Each individual topic has its own little focal points, whereby the fascinating is a particularly good match for the generous. As one can see, brown hues are used as rarely as grey and patinised colours. Even the fascinating does without excitement and echoism.

The template helps to determine a broad range of suitable nuances. Which colours from the colour studies were selected for the actual designs is and remains a question of customisable, creative and objectifiable rules. We present results and use images to demonstrate how neatly and easily recipes can be developed.

与前四组色彩相比，这四组色彩更淡些，但并不显朴素，只是不再有超强的简洁感和力量感。在该浅色调色卡中，蓝色和绿色色系居多，而红色调极少，易引起人们的共鸣。

每种主题下的配色方案都各有侧重点，但"引人入胜"色和"大方怡然"色又有些接近。仔细观察，我们不难发现棕色、灰色和铜绿色都很少出现。即使是"引人入胜"色系，也不会令人感到兴奋，更不会给人留下深刻的印象。

此配色板有助于人们选出恰当的色彩。这些经过色彩研究而挑出的色彩，既能直接应用到实际设计中，也能根据需求、创意和客观条件等进行个性化配色。所展示的这些方案和图片也充分表明，配色有时是十分灵活、简单的。

ELEGANT   优雅内敛

According to the Etymologisches Wörterbuch (Etymological Dictionary; dtv, 1997) by Wolfgang Pfeifer, if something is raised above the ordinary, is characterised by exceptional dignity and noble plainness, and is tasteful and elegant, then it is possible to incorporate the content of stateliness into creative design. Even a plain, circular element has this form of pleasant dignity based on experience – more than any little button or necklace: a smooth, white light switch.

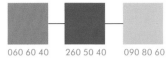

000 90 00   090 80 60   060 60 40

Generous ambiences with space on all sides are the basic prerequisite for unconfined living. In addition to interior-design rules relating to visually open spaces, distance, and acoustic, olfactory, colour, light-emitting, and pattern-related qualities, characteristics of skilfully designed parts of buildings, like lobbies, restaurants, terraces, bars, corridors and conference rooms play an important role and are formative of the guests' value parameters. The hotel rooms themselves represent the culmination of requirements and relations to reality. Small rooms, too, should be given special attention: elegant simplicity is better than even the smallest attempt at concealing decorative abundance.

080 70 60   080 90 20   060 60 40

A bright interior scene accompanied by a gentle red prompts ambivalence to the solemn hotel room designed in deeper shades. The hotel lobby fascinates an astonished audience. The bolide-like curves form an unusual roof in the hall. Such impressive dynamics are reminiscent of giant waves and bulging sails, of dreamlike, floating cloud formations and golden wings.

080 70 60   080 90 20   060 60 40

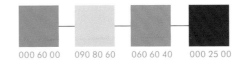

000 60 00   090 80 60   060 60 40   000 25 00

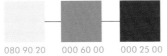

080 90 20   000 60 00   000 25 00

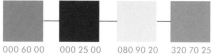

000 60 00   000 25 00   080 90 20   320 70 25

000 25 00   000 90 00   320 70 25   080 90 20

060 60 40   000 25 00   000 90 00   080 90 20

This is a space for conversations, in which to have a rest, to collect oneself and have a cup of tea; time to read the daily papers, the news, followed by the anticipation of a visit to the opera or a meeting with business friends or a reunion with a close friend one has missed for a long time.

根据Wolfgang Pfeifer编撰的Etymologisches Wörterbuch（词源字典；dtv，1997），如果一个物品并不平凡，而且显得高贵优雅，那么它身上的这种庄严便能加以创新。即使是简单的圆形物品，也会有种尊贵感，如小巧的纽扣或项链、平滑的白色灯开关。

高雅大气的氛围是自由生活的前提。室内空间应视野开阔，声效和气味良好，距离、色彩、灯光和图案应用到位；房屋各组成部分的设计也十分重要，如大厅、餐厅、露台、楼梯、会议室等，这些都是客人衡量酒店价值的标准。酒店客房本身就能反映现实社会以及人们的高要求。小房间设计更需要特别注意，保留空间的雅致与朴素，要好过有意去隐藏丰富的装饰。

亮堂的室内布景，加上些许柔和的红色作点缀，能提升酒店空间因深色背景而产生的沉重氛围。酒店大堂令人惊叹。火球般的曲面形成一个独特的屋顶，装饰着酒店大厅，十分迷人。其展现出来的动感更令人印象深刻，让人不禁想到翻滚的巨浪、鼓起的船帆，又似乎是梦幻般的云雾，或是金色羽翼。

在此交谈小憩、饮茶静心、阅报听剧、聚朋会友、促膝畅谈，是最理想不过的了。

230 80 15

060 60 40

080 90 20

000 90 00

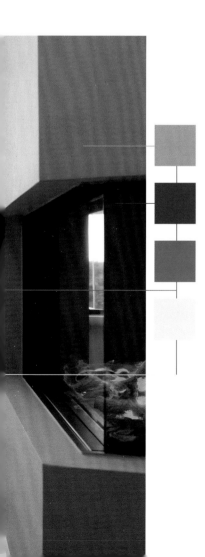

060 60 40

000 25 00

260 50 40

080 90 20

/ Chapter 1

PRINCELY

皇室隽永

The pie chart reveals that being princely implies a heavy burden. The substantial nuances point to that.

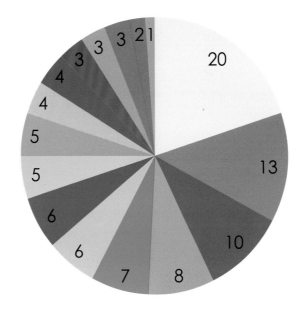

此扇形图中的各种色彩之间差别较小，这表明要打造隽永之风，并非易事。

The bar graph also demonstrates the cost-immanent consequences of practising design with princely colours. Understatement is inacceptable.

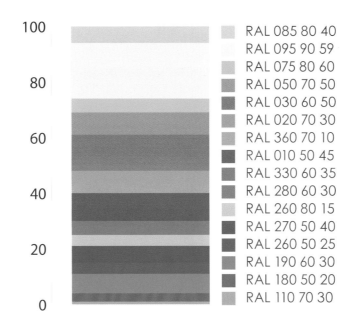

RAL 085 80 40
RAL 095 90 59
RAL 075 80 60
RAL 050 70 50
RAL 030 60 50
RAL 020 70 30
RAL 360 70 10
RAL 010 50 45
RAL 330 60 35
RAL 280 60 30
RAL 260 80 15
RAL 270 50 40
RAL 260 50 25
RAL 190 60 30
RAL 180 50 20
RAL 110 70 30

条状色卡表明，把"皇室隽永"色系用到实际设计中时，必然要花费一番功夫，倘若功夫不到位，就很难达到效果。

Suggested crowns, fur drawings and down-to-earth wave or bar graphics prevail, every now and then there are dot-shaped sequences.

皇冠、软毛、波浪、条状图案，不时还掺入了些近似圆形的斑点。

PRINCELY

皇室隽永

The colours of noble grace and the sovereign will to assert oneself. Purple, red and royal blue, tsar green and imperial gold colour the princely appearance perfectly. A hint of a princely hat serves to establish the truth in terms of graphics and colour.

代表高贵、优雅和皇权的色彩都富有表现力和征服力,如紫色、红色、品蓝色、沙皇绿以及皇室金,这些色彩能将"堂皇"外观表现得淋漓尽致。皇冠就体现了图形和色彩应有的取向。

/ Chapter 1

SOPHISTICATED

精致得体

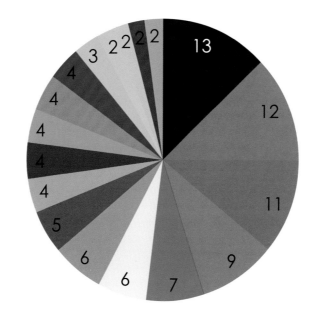

Purple and violet, which are frequently combined with black, occupy a prominent place in the sophisticated world. In most cases, amorphous flat shapes describe the nature of the sophisticated.

在"精致得体"色系中,深紫和淡紫所占比例最大,而且这两种色彩经常与黑色搭配使用。简单的不规则平面图形往往最能体现"精致得体"风范的本质。

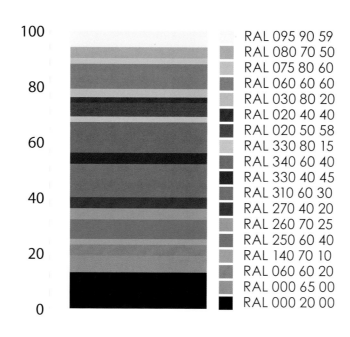

The scale is even more aristocratic than the originals on the right, because extravagant splendour accumulates, accompanied by plain satrap shades.

精致得体色彩的堆积,配上朴素而有力的色彩,使整个色卡中的色彩更具华贵气息,甚至超过了右侧的原色。

Lines and heavy monoliths are the core of the graphic content. Very rarely, small-scale interpretations can be made out.

这些图案主要以凝重的大色块和线条构成,并加入了些许淡描的细节。

SOPHISTICATED 精致得体

An evening gown for the festive gala – or does this represent the ambience of a Roman family claiming to belong to Caesar's relatives? The truth lies somewhere between imperial purple and haute couture.

这代表的是一袭华丽的晚宴礼服，还是凯撒大帝的罗马皇亲家族内部装饰？皇家御用的紫色和金色为此揭开了答案。

/ Chapter 1

NOBLE

隐性贵族

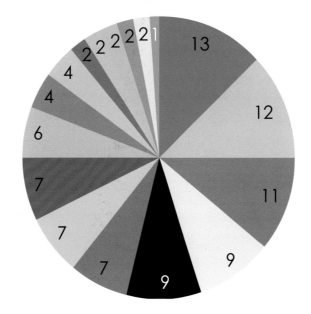

The colour circle shows more than a semblance of regal affectation. An air of tristesse, serenity and delicately powdered boredom wafts across the matt violet and oyster hues.

色盘中的色彩给人的不只是一种威风凛凛的表象。暗紫色和其他暗色表露出一丝忧郁和平静,还似乎有意掺杂了一种乏味感。

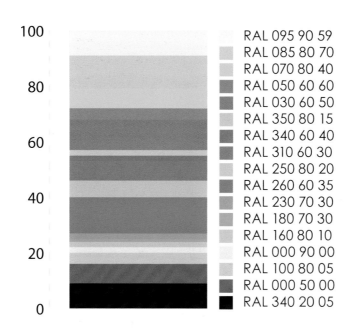

The bar chart also represents a harmonious colour ensemble. Beautifully unexcited and infectiously worthy of imitation.

图中的色彩显得十分协调,美观而沉静,渲染力强,值得借用。

RAL 095 90 59
RAL 085 80 70
RAL 070 80 40
RAL 050 60 60
RAL 030 60 50
RAL 350 80 15
RAL 340 60 40
RAL 310 60 30
RAL 250 80 20
RAL 260 60 35
RAL 230 70 30
RAL 180 70 30
RAL 160 80 10
RAL 000 90 00
RAL 100 80 05
RAL 000 50 00
RAL 340 20 05

Sparing use of signs: very plain, and well-rested, but distinctive in an extensively calm way.

这些图案单调简洁、有条不紊,而最打动人心的是一种沉着静默之感。

NOBLE

隐性贵族

Unexcited colours, the formal contents are similarly relaxed. The shades express the charm of silence and the hidden myth of modest but precious velvety dignity, as it could be admired in the swashbuckling films of the 1950s and 60s.

沉静的色彩、自由的形式，体现了"无声胜有声"的魅力和谦逊高贵背后的神秘感。这要是用在20世纪五六十年代的夸张电影中，会广受欢迎。

/ Chapter 1

SPLENDID

金碧辉煌

Winningly cheerful, that is how we experience the sunlit nuances. The watery blue hues also reveal shimmering surfaces.

在我们的意识中，灿烂的金黄色就是对胜利的欢呼。同样，用淡蓝色调也能绘出耀眼的外观。

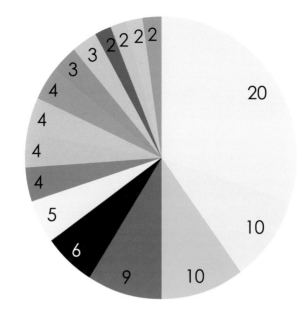

We perceive the sequence of shades as airy, playful, very friendly and sympathetic. There is not a hint of pompousness or blustering.

这些色彩给我们的感觉是轻快、俏皮、亲切、和谐，不带一丝傲慢和狂妄。

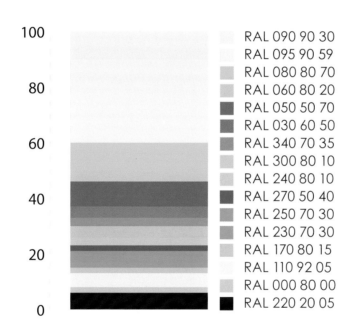

RAL 090 90 30
RAL 095 90 59
RAL 080 80 70
RAL 060 80 20
RAL 050 50 70
RAL 030 60 50
RAL 340 70 35
RAL 300 80 10
RAL 240 80 10
RAL 270 50 40
RAL 250 70 30
RAL 230 70 30
RAL 170 80 15
RAL 110 92 05
RAL 000 80 00
RAL 220 20 05

Semicircular hints of rainbows, yellow stars in the bright firmament and a black sparkler.

半环形状代表彩虹，黄色的星星在明亮的天空中闪烁着，还有一颗黑色的钻石。

SPLENDID 　　　　　　　　　　　　　　　　　金碧辉煌

Lots of bright shades shine towards us. Frequently, they are yellowish peach-coloured or bluish-rosé. Some interpreters saw the splendour in polished piano lacquer or a black diamond.

许多明亮的色彩在我们眼前闪耀着，其中淡黄桃色和偏蓝的玫瑰红色居多。有的解读者还能从亮钢琴烤漆色或菱形黑块中解读出辉煌之气。

/ Chapter 1

PRINCELY
皇室隽永

SOPHISTICATED
精致得体

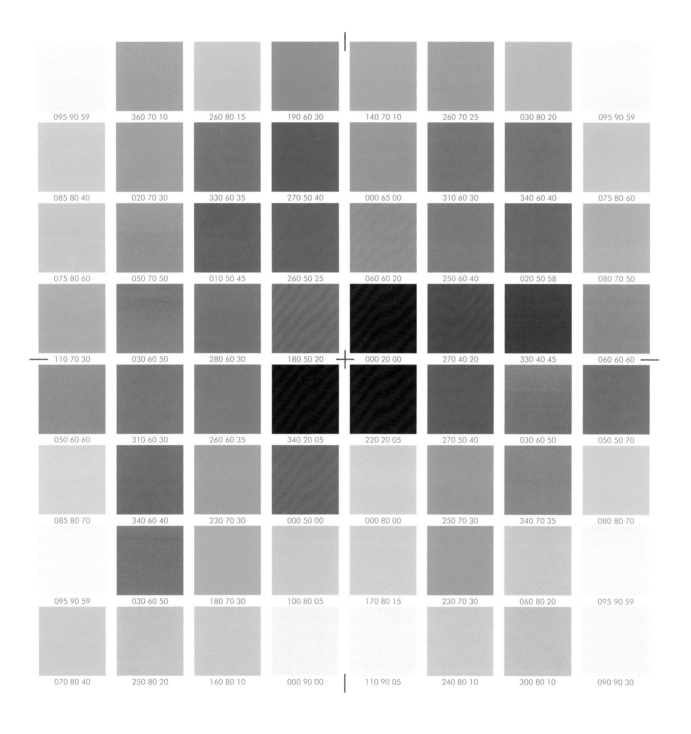

NOBLE
隐性贵族

SPLENDID
金碧辉煌

PRINCELY SOPHISTICATED NOBLE SPLENDID

皇室隽永
精致得体
隐性贵族
金碧辉煌

Colours that might have been borrowed from spectacular movies of the past, when actors performed with powdered wigs, pigtails, pinned-up hairdos, and artificial beauty spots. Quite rococo-like and playful, Mozart and Salieri, with love from Salzburg or Vienna or Paris. Petty princes and Louis XVI of France dance in silk costumes and brocaded gowns towards their downfall or the imminent loss of their power. But it was beautiful, cheerful and exciting, captivating and sensual all the same.

Lightness and a fresh expression characterise sunny and golden, shimmering colours. It was not until later that they became more sophisticated, bizarre and theatrical. Frequently, we discover such hotels in big, historic cities, in small castles in the countryside and in villas next to rivers, lakes and the sea.

或许这些色彩都来自以往场景华丽壮观的电影中，当时的演员头戴粉色的假发，拖着长长的辫子，或者顶着高高的发髻，甚至故意画上美人痣，十分有趣，像极了洛可可风格。不管是音乐家莫扎特和萨列里，还是来自萨尔斯堡、维也纳和巴黎的普通人，都曾喜欢这类色彩，如法国国王路易十六，这些卑微的王者们在面临失权和衰落时，还身着锦衣华服。这些色彩实在美丽动人，易令人目眩神迷。

耀目的金色奠定了整体清新明亮的风格，而越往中央，色彩越深，也显得更加迷离和奇异。这种设计风格通常会出现在历史名城或大都市的酒店中，也会出现在乡间的城堡和屹立于河边、池边或海边的别墅中。

SOPHISTICATED

精致得体

Truly sophisticated colours, which define the scenery with their paradisiacal and mythical violet-purple hues, help to make the guest's stay a delight. One might be listening to "The Bat" by Strauss and reading de Maupassant, looking at Klimt and Ensor or Ernst and de Lempicka. The sophisticated becomes tolerable for us from the cultural memory of freedom-loving times, which often reduced the prevailing conventions to absurdity through exaggeration. Taking another look at the colour range, we also recognise the paradox of the harmony of shades: both challenging, hierarchical, precious and noble, and mysteriously morbid and vulnerable.

That which is described as "sophisticated" takes place on a worldstage or is in the style of the big, wide world. The sophisticated is not classless, but

330 40 45   310 60 30   020 50 58

multi-cultural; whether turban or sarong, dinner jacket or prêt-à-porter, the main thing is to be chic, charming and beautiful. "Sophistication" is still associated with images reminiscent of the belle époque at the turn of the nineteenth century and the Art Deco of the 1920s: long cigarette tips, seductive bobs, spaghetti lamps, Charleston and early jazz.

Right in the centre of big cities, in the countryside, in the mountains and on the shores of lakes and oceans one finds hotels that have the exclusive flair of nostalgia and internationality, of subtlety and elegant temptation. Especially in these times of high-tech and the virtual, the ambience created by back-stitched surfaces, dark polished wood, classic brocade fabrics and high beds meets with a growing number of enthused guests.

　　真正精致的色彩，如用神秘而妙不可言的紫色调定义酒店布景，有助于客人愉快地逗留。人们可以听着施特劳斯的"The Bat"，读着莫泊桑的小说，看着克里姆特、恩索尔或德·蓝碧嘉的画作。别致使得我们从自身的文化记忆转向热爱自由的时代的文化记忆，它往往通过夸张的手法摆脱惯例，使其显得荒诞。再看看颜色范围，我们也认可色调和谐中的矛盾：充满挑战且层次分明，稀少及高贵，带点神秘的忧郁及脆弱都交融在一起。

310 60 30   020 50 58

000 65 00   000 20 00   060 60 20

000 65 00   020 50 58   000 20 00

无论是在大都市的中心或是在乡村、山区、湖畔或海岸边，都有一些风格独特、怀旧而国际范的酒店，它们优雅迷人。尤其在这个日趋高科技的虚拟化时代，背缝面、深色抛光的木材、经典锦缎面料和高床营造出的氛围吸引了越来越多饶有兴趣的客人。

000 20 00   020 50 58

被描述为"精致得体"，则意味着受到世界广泛认可或契合大千世界的时尚。精致得体的风格并非无阶级的，但它意味着多种文化的融合；不管是穆斯林的头巾或马来西亚、印度尼西亚人穿的沙笼，不管是晚礼服或成衣，它们都非常时尚、迷人和美丽。"别致"而今仍与20世纪之交那个歌舞升平年代的形象和20世纪二十年代的装饰艺术相联，如长烟嘴、诱人的短发、绝缘套管灯、查尔斯顿舞和早期的爵士乐。

| 000 65 00 | 020 50 58 | 000 20 00 |

| 060 60 20 |
| 000 65 00 |
| 000 20 00 |
| 020 50 58 |

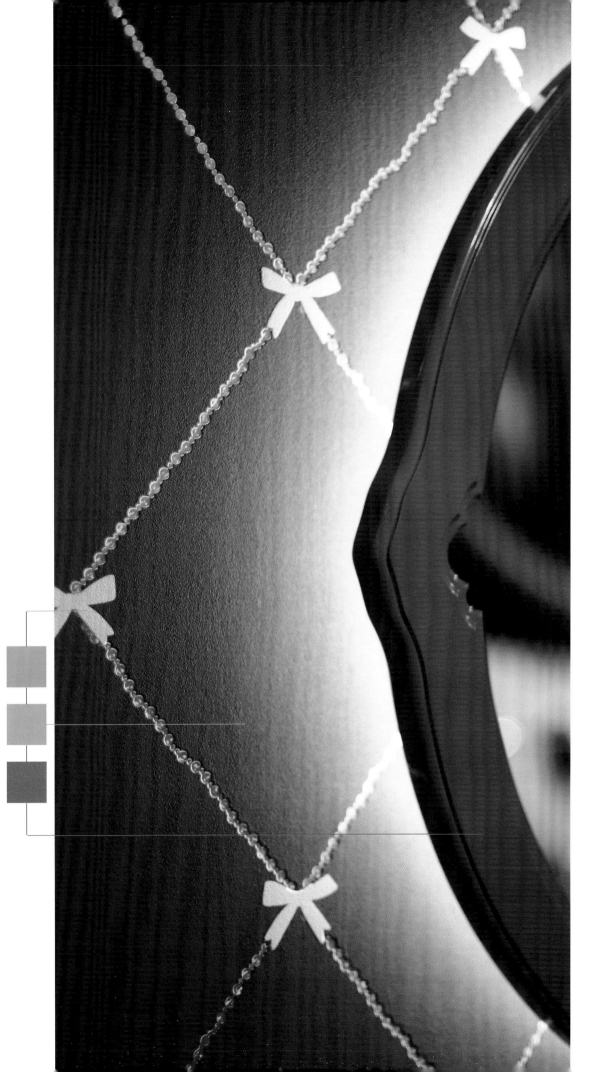

| 060 60 20 |
| 000 65 00 |
| 020 50 58 |

/ Chapter 1

BOURGEOIS

中庸务实

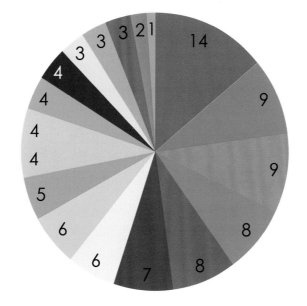

Like a walk through a furniture shop: the demonstration of the well-behaved average comes to life. In between times, the danger of falling asleep from exhaustion on a stray sofa is omnipresent.

色盘为人们带来仿佛在家具店中散步的感觉。平日里中规中矩、不出挑的色彩此时充满了生命力。疲惫的人们时不时会产生在家具店一角散落的沙发上睡着的冲动。

More warming than cooling, more standstill than movement, more present than future. Each shade is set to "better safe than sorry".

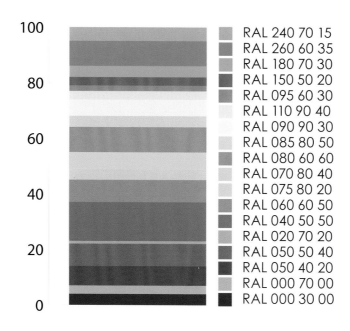

更暖而不是更冷，更趋向静止而不是运动，更关注现在而不是未来，每个色调都趋向于保守。

The bourgeois is neither round nor playful, but down-to-earth, stable and straight.

中产阶级既不圆滑也不热衷玩儿，他们脚踏实地、坦诚务实，稳重且正统。

BOURGEOIS 　　　　　　　　中庸务实

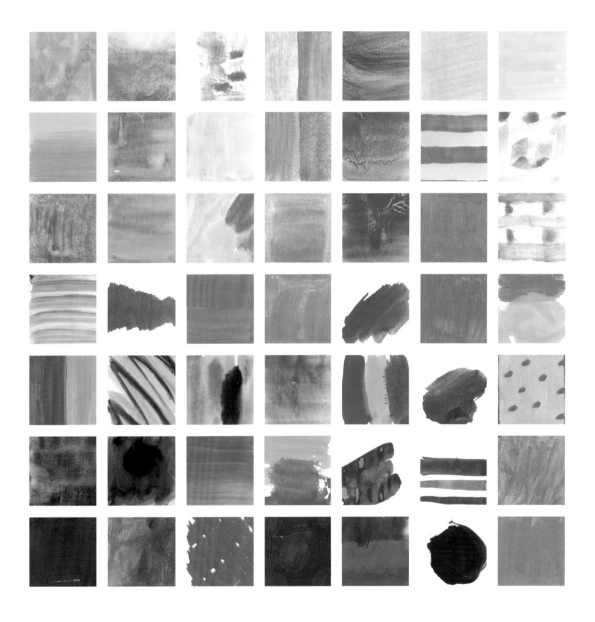

The colour spectrum reveals a quasi modern approach. Maximum-benefit, blunt mid-range hues, a few fresh nuances of blue, green and yellow, and about one third of limestone and pastel colours round off the colour spectrum.

色谱揭示了准现代的手法，包含了经过优化的、中庸而柔和的色调，以及清新的蓝色、绿色和黄色，而石灰色和其他柔和的色彩占据了约三分之一的色谱。

/ Chapter 1

STYLISH

时尚现代

For the most part, the shades give a sober, subject-related impression. Constructivist and zeitgeisty-futuristic ideas have accompanied the participants in our study.

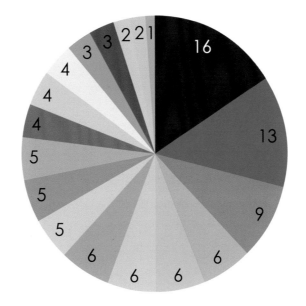

在大多数情况下，此色调给人留下主题清晰明确的印象。在研究中，我们还发现此色调常与结构主义和未来主义相联系。

The colour stripes show subject-related backgrounds. For this reason, they do not appear particularly lively but decidedly functional.

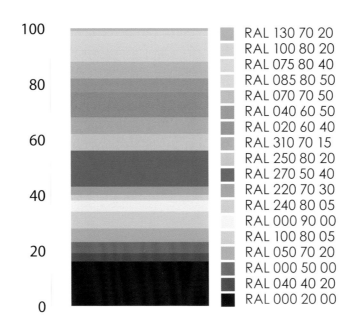

RAL 130 70 20
RAL 100 80 20
RAL 075 80 40
RAL 085 80 50
RAL 070 70 50
RAL 040 60 50
RAL 020 60 40
RAL 310 70 15
RAL 250 80 20
RAL 270 50 40
RAL 220 70 30
RAL 240 80 05
RAL 000 90 00
RAL 100 80 05
RAL 050 70 20
RAL 000 50 00
RAL 040 40 20
RAL 000 20 00

色条可显示与主题相关的背景。出于这个原因，它们不需要看起来很明快，但具有很强的实用性。

The shape-forming representations also define a graphic, boxy vocabulary. Only once or twice, circles and suggested waves can be recognised.

它们往往构成几何、方正的图形。圆圈和波浪形只是偶尔出现的情况。

STYLISH 时尚现代

The colours and shapes of the interpreters can be described as concrete. No flourish, no tasteless arabesque adorns the somewhat analytical representations. Seemingly, associations to past stylistic eras no longer matter.

图中所诠释的色彩和形状就如混凝土一般，看起来并不耀眼，也没有品味低俗的蔓藤花纹装饰。看来向过去的时代风潮致敬已无必要。

/ Chapter 1

UPSCALE

卓越上层

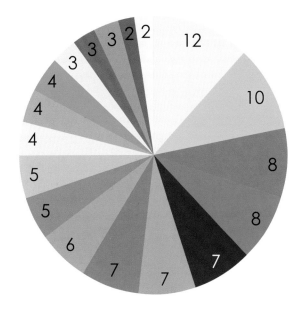

The result looks cheerful and buoyant, despite a small age percentage of black. Almost all shades are pure and unblended.

尽管含有小部分黑色，看起来仍欢快活泼。几乎所有的色调都很纯粹，而非混合而成。

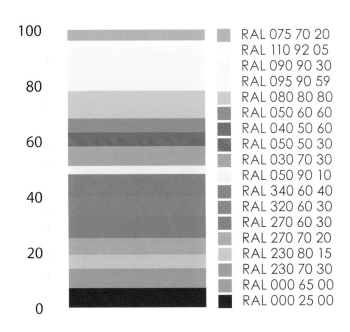

RAL 075 70 20	
RAL 110 92 05	
RAL 090 90 30	
RAL 095 90 59	
RAL 080 80 80	
RAL 050 60 60	
RAL 040 50 60	
RAL 050 50 30	
RAL 030 70 30	
RAL 050 90 10	
RAL 340 60 40	
RAL 320 60 30	
RAL 270 60 30	
RAL 270 70 20	
RAL 230 80 15	
RAL 230 70 30	
RAL 000 65 00	
RAL 000 25 00	

The bar diagram displays a high degree of attractiveness. It displays many features of high-quality silky colourfulness and thus exudes a textile charm.

条形图展示了极强的吸引力，它具有柔滑而高品质的色彩特征，因而散发出优质纺织物才有的魅力。

For their representation the painters were content with nothing but hints of abstractions, frequently graphical and with visible brushstrokes.

画者在绘图过程中只需绘制或抽象或明了的图案，而后者占多数。

UPSCALE 卓越上层

The colour range is splendid and fashionable. It cannot be denied that the colours are elegant and display a good deal of sophisticated chic. Nonetheless, they tend strongly towards timelessness.

此色域中的色彩明艳时尚。不能否认的是，这些色彩非常典雅，而且精致时尚，且趋于经久不衰。

/ Chapter 1

CONSERVATIVE

保守克制

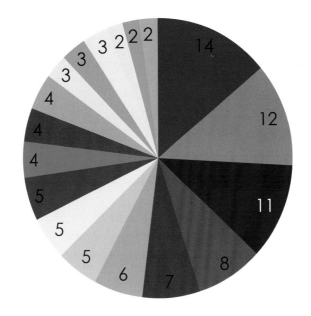

The colour circle conveys the impression of woody immobility and the absence of future. The readiness to show one's colours is not pronounced.

色盘传递了木头似的静止和看不到未来的感觉，色彩并不明确。

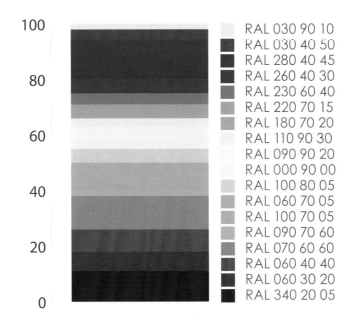

The heavy brown fills the scale with sounds like a thundering echo. At the very end there is a little more bordeaux-red. All in all, this is not a very friendly colour scheme.

由深褐色填充的部分，给人带来雷鸣般的感觉；最后一小部分出现了波尔多红。总的来说，这并非一个出色的配色方案。

Any round, formal content is missing. On the contrary, plenty of angular, diagonal, intersecting, and martial shapes.

图中不见任何圆形或规规矩矩的形状，而出现了很多三角形、斜线、交叉以及富有力量感的形状。

CONSERVATIVE 保守克制

The conservative shuns bright light. This term does without rosé and yellow, red and reseda, spring green and orange. The deep watery blue restores hope for future viability.

保守风格的色调会回避明亮的光线，之中并无玫瑰红色、黄色、红色、浅绿色、嫩绿色和橙色。唯有水汪汪的深蓝色带来些许生机。

/ Chapter 1

BOURGEOIS
中庸务实

STYLISH
时尚现代

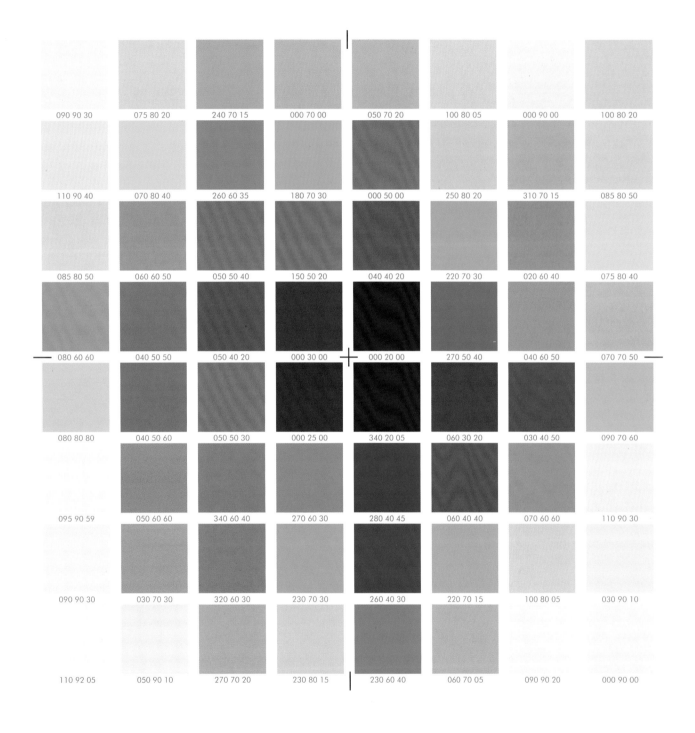

UPSCALE
卓越上层

CONSERVATIVE
保守克制

BOURGEOIS
STYLISH
UPSCALE
CONSERVATIVE

中庸务实
时尚现代
卓越上层
保守克制

Bourgeois colour virtues shape the colouring of the entire spectrum. It is composed of shades that are substantial and able to express an undeniable, conservative approach to values and the world. A glamorous event does not take centre stage as a signal image. Instead, distinctive feel-good elements enter the awareness of the beholder.

The large number of wood, cognac, terracotta and clay colours, as well as dark anthracite and slate hues in addition to the much warmer cream nuances determine the expressiveness of the cosy shared colour type that invites us to linger. Designing with these shades leads to a high level of general acceptance because they transmit an unmistakeable sense of safety. One side effect is that people are given a comforting feeling of security and stability.

  中庸务实的色彩构成了整个色谱，它由大量可明显体现中产阶级保守的价值观和世界观的色调组成。鲜艳夺目的色彩不会占据中心位置，相反，比较舒适的色彩元素常常映入眼帘。

  大量的木质色彩、琥珀色、陶色、泥色、黑煤色和深蓝灰色，以及温暖的奶油色使得此风格色调看起来舒适、宜人，仿佛在邀请客人多加逗留。使用此色调设计的作品传递了坚实的安全感，常给人们带来安全、稳定的感觉，易于被大众接受。

STYLISH 时尚现代

Gentle, noble and enigmatic woody colourfulness provides the generously designed ambience with an impressive backdrop. Even the circular bath tub shines in deep brown below the circular canopy suggesting illuminated tropical rain. One cannot easily forget such designs; their narrative message is very precious. Hotel architecture and interior design in particular are comparable with stage design. The scenic message determines whether one likes it or not.

lighting is of tremendous importance; changing, mysterious light impressions can be created using grazing light and selectively positioned spotlights. Upper-class glamour and generous dimensions belong to the stylish life, as do really good service and top-quality acoustics, ergonomics and designs using colours that portray a certain international chic and are made of high-quality materials.

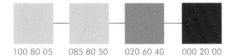
100 80 05   085 80 50   020 60 40   000 20 00

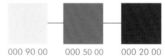
000 90 00   000 50 00   000 20 00

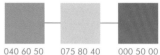
040 60 50   075 80 40   000 50 00

温和、高贵而神秘的木质色彩为大气的设计氛围提供了令人印象深刻的背景。圆形天蓬下的深棕色圆形浴缸使人想起热带亮晶晶的雨滴。它们的设计语言如此罕见，人们将难以忘记这样的设计。酒店建筑，特别是室内设计可与舞台设计相媲美，舞台氛围便可决定人们喜欢与否。

照明极其重要，精心设计的聚光灯可营造出变化、神秘的灯光。时尚生活还包含了宽敞的空间、周到的服务、顶级的传声效果和工效学效果。此外，设计还采用了可反映一定国际时尚的色彩以及优质材料。

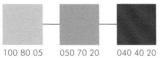
100 80 05   050 70 20   040 40 20

000 90 00   040 40 20   270 50 40   310 70 15

050 70 20   000 90 00   000 20 00

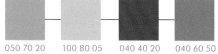

050 70 20   100 80 05   040 40 20   040 60 50

070 70 50   075 80 40   000 20 00

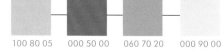

100 80 05   000 50 00   060 70 20   000 90 00

The results are certainly impressive because one realises that they are all of a piece and thus offer plausibility and design orientation. Our feelings about this interior-design vocabulary resemble our feelings about very exotic, folkloristic or experimental examples: depending on our mood and whim, we like them without prejudice and enjoy their comfort.

成果自然令人印象深刻，因为设计师意识到它们都是一体的，这为设计提供了合理性和方向。我们对室内设计语言的感受非常类似于对异国风情、民俗和试验实例的感受。所使用的语言取决于我们的情绪，有时纯粹是心血来潮。我们喜爱它们，不带任何成见，同时享受它们带来的愉悦。

000 20 00

040 40 40

050 70 20

020 60 40

040 60 50

075 80 40

240 80 05

050 70 20

000 20 00

070 70 50

STYLISH

/ Chapter 1

PREMIUM

超凡优选

The colours look as if they come from the repertoire of ambitious designers, who want to ennoble a bourgeois ambience with royal grandeur. Maybe many a hotel lover is aken by the pretty appearance.

此风格的色彩看起来，仿佛来自雄心勃勃的设计师的保留项目，它试图为中产阶级的氛围添加些许皇家气息。许多酒店爱好者也许会为此漂亮的外观所吸引。

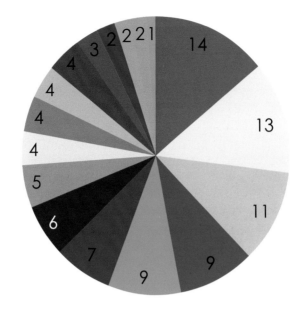

In all innocence, the stripes demonstrate the always successful recipe of gold, purple and royal blue.

此条形图已不知不觉地成为了金色、紫色和深蓝色的配色宝典。

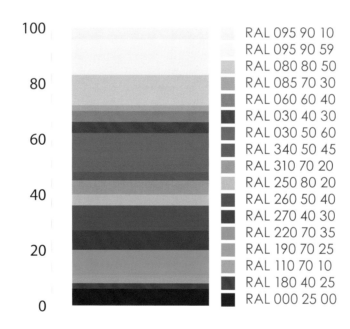

The symbolic power of "a" and "1" suffice for the visualisation of the demand.

"a" 和 "1" 符号的力量足以满足可视化的需求。

PREMIUM   超凡优选

Classifying something as "premium" reveals latent doubts, just like the loud proclamation of placing "complete confidence" in something. Maybe premium has less class than the more modest "exceptionally good" or "sublime".

如把某物归类为"超凡优选"类别,即代表着我们心存怀疑,就像我们大声说出对某物"很有信心"的感觉一样。也许"超凡优选"这个词与"好得不得了"和"绝妙"相比,还是差了些。

/ Chapter 1

WORLD-CLASS

世界一流

The manifestations of this ambience are statements of eternal prestige. They preach status and image, which reflect back on the guests.

此种格调彰显了酒店长期以来建立的声誉，宣扬了其地位和形象，同时对客人也有所影响。

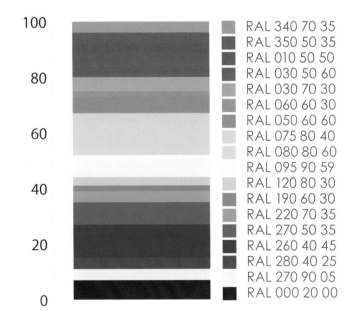

RAL 340 70 35
RAL 350 50 35
RAL 010 50 50
RAL 030 50 60
RAL 030 70 30
RAL 060 60 30
RAL 050 60 60
RAL 075 80 40
RAL 080 80 60
RAL 095 90 59
RAL 120 80 30
RAL 190 60 30
RAL 220 70 35
RAL 270 50 35
RAL 260 40 45
RAL 280 40 25
RAL 270 90 05
RAL 000 20 00

The colour sequence can only imperfectly describe the shining splendour of today and yesterday. Sometimes, the most beautiful place can be found in the world's most gorgeous lobbies between palm trees, plush and piano.

色彩的顺序只能不完整地描述过去和现在的辉煌。有时候，世界上最美的地方是闪耀于棕榈树、长毛绒和钢琴之间的最华丽的酒店大堂。

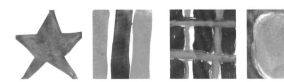

The interpreters forewent delicate decoration. In order to suggest world-class, powerful onomatopoetic appeals are required.

诠释者放弃了精致的装饰。为了营造世界一流的感觉，有强烈吸引力的描绘手法是必需的。

WORLD-CLASS 世界一流

The "world-class colours" are reminiscent of the colourful garb of world-class athletes. It signalises to fans that world-class results can seldom be achieved by the colourless.

"世界一流的颜色"让人联想到世界一流运动员多彩的装束。它欲告诉痴迷者们一流的配色成果很少能够由平淡的颜色实现。

/ Chapter 1

GARISH

About eighty percent of the shades are located in the warm-toned area. We can see that with green and Prussian blue alone the world is lacking in libidinous effectiveness.

炫丽夺目

大约80%的色调位于暖色调区域。我们可以看到，单纯的绿色和铁蓝色缺乏春意盎然的感觉。

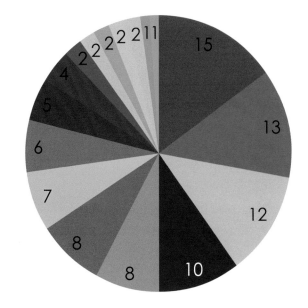

The pretence of a gala evening or celebratory banquet with the "queens" of society requires shades such as these. Those who do not have them strive in vain for recognition and to avoid being forgotten.

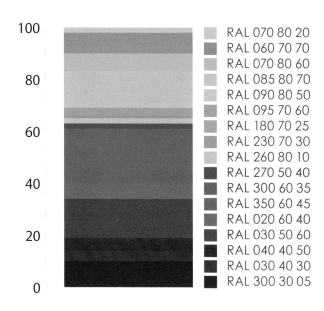

上层社会矫揉造作的盛大晚会或庆祝宴会需要这样的色调，它们对获得认可和避免被人遗忘的努力颇有成效。

Stripes, whorls and circles suffice as formal content of the garish. The materiality is compelling enough.

醒目的形状只需条带、旋涡和圆圈便已足够，它的物质特征已十分引人注目。

GARISH　　　　　　　　　　　　　　炫丽夺目

Violet and pink hues, which are accompanied by orange and sunny yellow, give the strongest impression. Showing off a little gives people great pleasure; enjoyment and signal effect are congruent.

紫罗兰色调和粉红色色调，以及橙色和日光黄色，令人印象深刻。些许炫耀能为人们带来快乐，图中符号传递的感受和带来的享受相一致。

/ Chapter 1

DANDIFIED

狂欢浮华

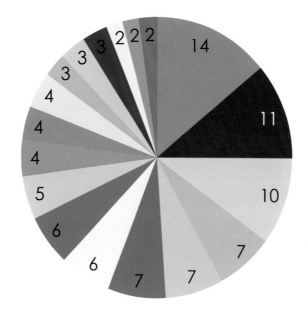

A beautiful pie chart has been created. Its appeal arises from the slightly wild, high-contrast colour combination. For example, when yellow-gold and pink encounter black.

这是一个美丽的饼形图。它的魅力源于略带野性和高对比度的颜色组合。例如，金黄色和粉红色与黑色相遇。

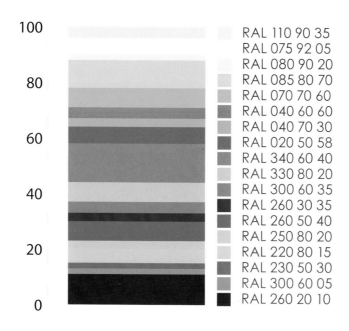

The well-ordered colour sequence is less amusing than the combination in the pie chart that is defined percentage-wise as a majority.

与百分比明确的饼形图相比，井然有序的颜色序列较无趣。

RAL 110 90 35
RAL 075 92 05
RAL 080 90 20
RAL 085 80 70
RAL 070 70 60
RAL 040 60 60
RAL 040 70 30
RAL 020 50 58
RAL 340 60 40
RAL 330 80 20
RAL 300 60 35
RAL 260 30 35
RAL 260 50 40
RAL 250 80 20
RAL 220 80 15
RAL 230 50 30
RAL 300 60 05
RAL 260 20 10

Little stars, sharp angles, suns and soft diagonals make up the main part of the distinctive, symbol-like drawings.

小星星、尖角、太阳和柔和的对角线充当了这些符号般的、独特的图画的主要部分。

DANDIFIED   狂欢浮华

Have the shades of colourful excesses or of the carnival in Rio or of a quirky ambience in the night-life district of a seaport served as a model? Colours make many things more beautiful than beautiful.

难道这些反映里约狂欢、海港上典型夜生活的光怪陆离氛围的鲜艳色彩不能作为设计参考？色彩总能使事物愈加美丽。

/ Chapter 1

PREMIUM
超凡优选

WORLD-CLASS
世界一流

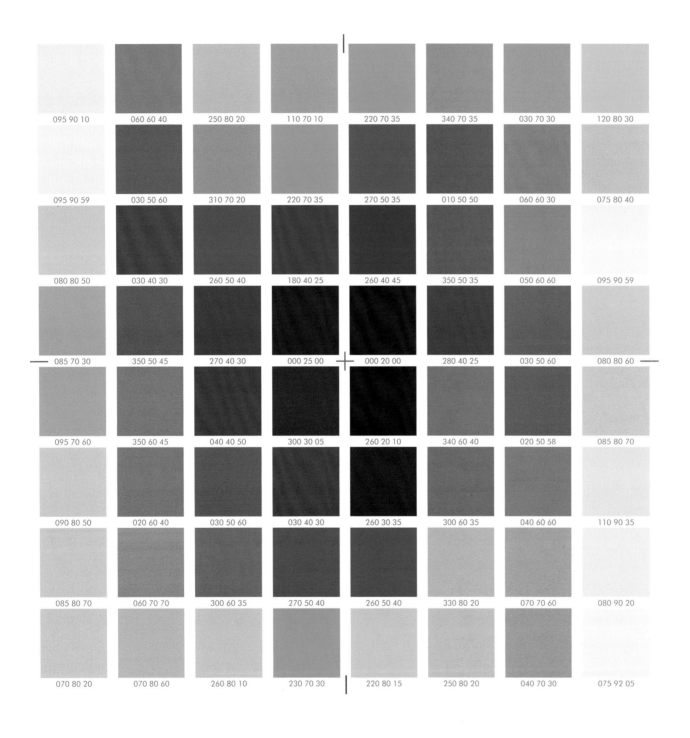

GARISH
炫丽夺目

DANDIFIED
狂欢浮华

PREMIUM
WORLD-CLASS
GARISH
DANDIFIED

超凡优选
世界一流
炫丽夺目
狂欢浮华

They exist all over the world: hotels with colourful and glossy status symbols. Hardly anyone can avoid feeling impressed by their glamour. Expensive shades rich in tradition like Rose Madder, yellow gold, jade green or the elitist caput mortuum (a dull, grey dark red) and brown violet, as well as deep black, exist on this scale in deep matt, in shiny super glossiness and with a metallic surface shimmer.

Of course ambiences designed in such a way frequently have an awe-inspiring effect. For the benefit of other guests they silence voices that are otherwise too loud. One can see that colours can also demonstrate pedagogical skills. The sixty-four shades on the opposite page can filter out differentiated and combined themes. For the following example of a hotel, we have focused on "world-class" content.

世界各地，拥有光鲜地位的酒店无处不在，几乎没有人不为它们的魅力所折服。传统高贵的色调，如玫瑰茜草色、金黄色、翡翠绿、印度红（一种沉闷、带灰色的暗红色）、褐紫色以及深黑色，经亚光处理后与其他泛着金属光泽的材质相映衬。

当然，以这种方式营造的氛围有着令人肃然起敬的效果。相比其他色调，它们时常出现。人们可以看到，色彩也能起到教育示范作用。上一页的64种色调可滤除分化和组合的主题。接下来的内容中，我们举一个"世界一流"的例子。

WORLD-CLASS   世界一流

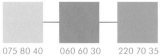

Across the boundaries of nations and tastes, interior design must provide people with a sense of security, freedom and congeniality. In doing so, it is always sensible to furnish the universally valid with folkloristic features, if only in the form of a view of a historic landmark, such as Cologne Cathedral.

The rooms feature different temperaments in their colour schemes and furnishings. Fabrics by the windows, on the beds and the floor ensure atmosphere and continuity given the right dosage of microclimate. Natural, faunistic and floral textiles in particular increase the air humidity. All wool products and cotton fibres are especially suitable for this purpose.

Large pictures on walls in a wide variety of colouristic and graphic facets are much in demand, and so are textile floor products designed in the same way. It is also clear that strip lights, indirect lighting and standard and table lamps can create a lively interior appearance provided that they are skilfully combined. Even in smaller rooms, we appreciate it when there are at least three other light sources in addition to the main light on the ceiling. Examples illustrate this time and again: Mobile light sources are not necessary for practical reasons, but they offer the hotel guest individual adaptability. Light is located where the guest wishes to have it.

跨越了各国和各种美的界限，室内设计必须为人们带来安全和自由感，引起人们的共鸣。在此过程中，装潢布置从大众熟悉的民俗等出发是明智的，如科隆大教堂等历史地标的挂画等。

客房由于不同的色彩方案和布置拥有不同的特点。窗户边、床上和地上的面料可确保室内温度和氛围。特别是天然的、与动物相关的和印花的纺织品可增加空气湿度。出于此目的，所有羊毛产品和棉质布料异常合适。

墙上多幅丰富多彩的大型挂画大受欢迎，图案色彩多种多样的地毯、垫子等也如此。显然，条形灯、间接照明、落地灯和台灯等巧妙结合的话，可以打造出生动活泼的室内。即使在较小的房间，室内除了天花板上的主照明以外还有其他三种照明，也是我们所赞赏的做法。这种例子层出不穷。移动光源实际来说并无必要，但它们为酒店顾客提供了一个适应的过程。灯应布置在客人有需要的地方。

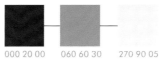

000 20 00   060 60 30   270 90 05

000 20 00   075 80 40   270 90 05

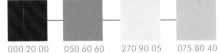

000 20 00   050 60 60   270 90 05   075 80 40

030 70 30   120 80 30   050 60 60   190 60 30   270 50 35

Even for seasoned business travellers, life far away from home is not always a pleasure. However, even the smallest of gestures can help to alleviate stress and thus counteract the sense of loneliness.

If the hotel is located in the city centre, information about cultural events, cinema programmes and live music performances, ideally within walking distance, are appreciated, as are city guidebooks and the current issue of the local newspaper. The popular small bowl of fruit, a little bottle of sparkling water, snacks (but definitely no gummy bears!) and a little bouquet or an entertaining short novel, an exciting novella or a small book of poems or aphorisms by Rilke, Fontane, Kraus, Kleist, Wilde, et al. are also gratefully accepted.

075 80 40

000 20 00

270 90 05

075 80 40

030 50 60

000 20 00

270 90 05

090 80 60

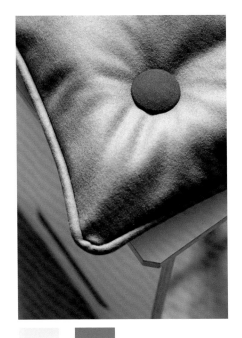

对经验丰富的商务旅客来说，远离家乡的生活并不总是一种享受。即使是最小的关怀也有助于客人舒缓压力，消除孤独感。

如果酒店位于市中心，店内提供城市指南和当地报纸，其刊载附近相关的文化活动、电影节目和现场音乐会等信息，客人会不胜感激。流行的小盘水果、小瓶苏打水、零食（当然不是小熊软糖！）、小花束、有趣的短篇小说、中篇小说，或者Rilke、Fontane、Kraus、Kleist、Wilde等人的小本诗集或警句也会为客人欣然接受。

270 90 05   030 50 60

060 60 30
030 50 60
350 50 35
000 20 00
270 90 05

060 60 30   270 90 05   030 50 60

/ Chapter 1

ENRICHING

丰富多彩

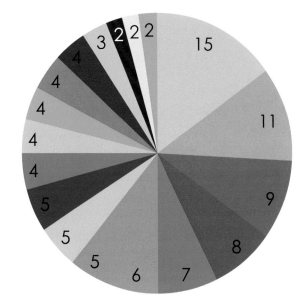

The round disc seems slightly ambivalent to the beholder. Violet and warm cognac shades rarely go well together, and nor do gentle yellow-green and ocean blue.

旁人看来，该色盘略显相克，紫罗兰色和暖暖的琥珀色并不匹配，柔和的黄绿色和海蓝色也很不搭。

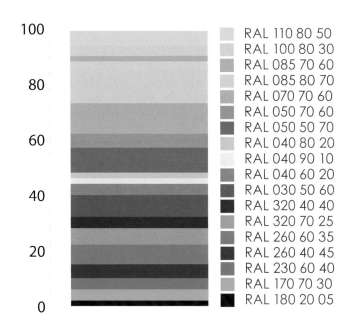

Though weaker than the colour circle, the stripe composition bathes what previously appeared contrary in a visible, soft light.

条形的色彩看起来虽比色环弱，但之前相克的颜色在这里看起来柔和且和谐。

RAL 110 80 50
RAL 100 80 30
RAL 085 70 60
RAL 085 80 70
RAL 070 70 60
RAL 050 70 60
RAL 050 50 70
RAL 040 80 20
RAL 040 90 10
RAL 040 60 20
RAL 030 50 60
RAL 320 40 40
RAL 320 70 25
RAL 260 60 35
RAL 260 40 45
RAL 230 60 40
RAL 170 70 30
RAL 180 20 05

Few symbols, here and there stripy or small circular shapes. Many amorphous forms are very close to plain flatness.

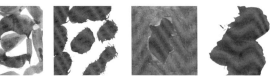

极少出现条形或圆形符号，许多都没有固定形状，极接近平面。

ENRICHING 丰富多彩

The quality of content is characterised by an explicitness that has little of the spectacular about it. Objectivity and function, which seem to be based on reason and honesty, are the focus of the painters as they interpret.

此风格贵在明确而不盛大。作此画者在诠释时，将焦点放在基于理性和诚实的客观性和功能性上。

/ Chapter 1

TOP-NOTCH

首屈一指

Purple, magenta, orange-red, discerning nuances of blue and gold provide the foundation for peak performances.

紫色、紫红色、橙红色、略有差别的蓝色和金色奠定了顶尖色调的基础。

The horizontal lines and bars illustrate the onomatopoetic but also the rare substance of the selected colour range.

横线和条状图阐明了此风格色彩的效果和实质。

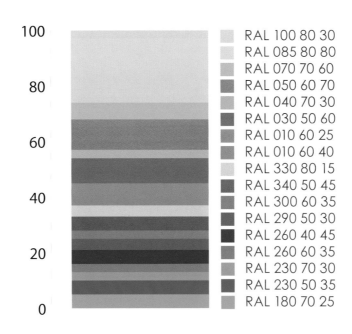

RAL 100 80 30
RAL 085 80 80
RAL 070 70 60
RAL 050 60 70
RAL 040 70 30
RAL 030 50 60
RAL 010 60 25
RAL 010 60 40
RAL 330 80 15
RAL 340 50 45
RAL 300 60 35
RAL 290 50 30
RAL 260 40 45
RAL 260 60 35
RAL 230 70 30
RAL 230 50 35
RAL 180 70 25

The pointed symbols signalise the purest meaning of the term.

尖尖的符号清楚地表明了"首屈一指"这个词的本意。

TOP-NOTCH　　　　　　　　　首屈一指

The demands made of on the top-notch deserve attention! Here, concrete-like reduction is obsolete. Bleak, whitewashed surfaces or the excesses of Alpine cosiness won't do either. Designing the colourful with flying colours works perfectly!

"首屈一指"的风格值得关注！混凝土质感递减已过时，萧瑟感、表面粉刷感或过多的山林美感也很少出现。色彩飞扬的设计才能出彩！

/ Chapter 1

DISCERNING

谨慎挑剔

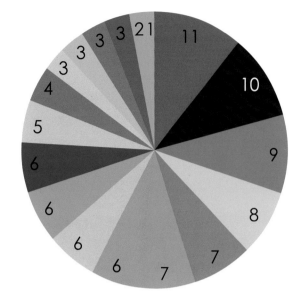

The colour shades are cool and pure. As we can see, warm nuances of brown and beige to umbra and green are not in demand.

此色调冷静、纯粹。正如我们看到的，有着细微差别的棕色和米色占据较大一部分而绿色所占很少。

Festive activities inspire the colour harmony represented here. It works with operatic settings and with scenes set in elegantly draped halls of mirrors lit with chandeliers.

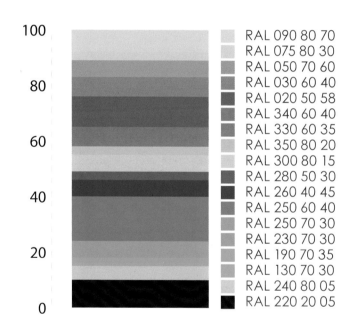

RAL 090 80 70
RAL 075 80 30
RAL 050 70 60
RAL 030 60 40
RAL 020 50 58
RAL 340 60 40
RAL 330 60 35
RAL 350 80 20
RAL 300 80 15
RAL 280 50 30
RAL 260 40 45
RAL 250 60 40
RAL 250 70 30
RAL 230 70 30
RAL 190 70 35
RAL 130 70 30
RAL 240 80 05
RAL 220 20 05

此风格的色彩在节庆活动中看起来颇为和谐。它可用于歌剧背景设置，与帷幔装饰、雅致的殿堂、镜面和吊灯相衬。

Strong signs of diagonals, grids and waves, with frequent pictorial lines running horizontally.

对角线、网格、波浪和横线频繁出现。

DISCERNING 谨慎挑剔

The shades of the discerning range between blue and red tones with various elements of brightness features and quality characteristics. Several shades of yellow make up the yellow-golden decoration, and black adds the necessary contrast.

这些色调介于蓝色和红色色调之间,有着不同的亮度特征与特性。几种黄色的色调构成了装饰性的金黄色,黑色则带来必要的反差。

/ Chapter 1

EXPENSIVE

贵气逼人

Colours like the medals or the badge of honour of any potentate or like the packaging of an outrageously expensive perfume or the entrée to the favourite hotel. Such a colour range never fails to arouse the desired effect.

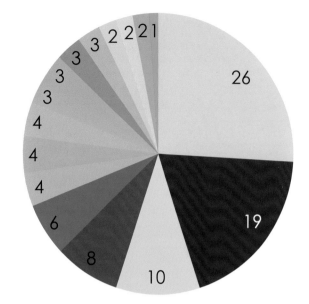

贵气逼人风格的色彩，如奖牌或当权者所授荣誉奖章的颜色，或见于贵得离谱的香水的包装，或见于最爱的酒店的入口。此色彩系列往往能达到所期望的效果。

That's what the design for a silk fabric should look like. If light falls on the folds and pleats, the silky shimmer will gleam like mother-of-pearl.

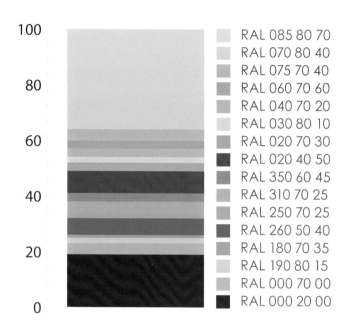

当光线落于面料的折叠处和褶皱上时，微光闪烁，丝绸面料的设计应该是这样的。

The massive black plain-coloured surfaces and bulky slightly rounded shapes are the foundation of the truly precious.

大面积的黑色、素净的表面和庞大、略显圆润的形状方显珍贵。

EXPENSIVE

贵气逼人

The shades with historic connotations bear the signs of what has an intrinsic value: yellow equals gold, red equals purple, violet equals ecclesiastical regalia, black equals power, dignity, sovereignty and authority.

有些色调有其历史内涵，承载了某些内在价值，如黄色等同于金色，红色等同于紫色，紫罗兰色象征着教会的权力，黑色意味着权力、尊严、主权和权威。

/ Chapter 1

ENRICHING
丰富多彩

TOP-NOTCH
首屈一指

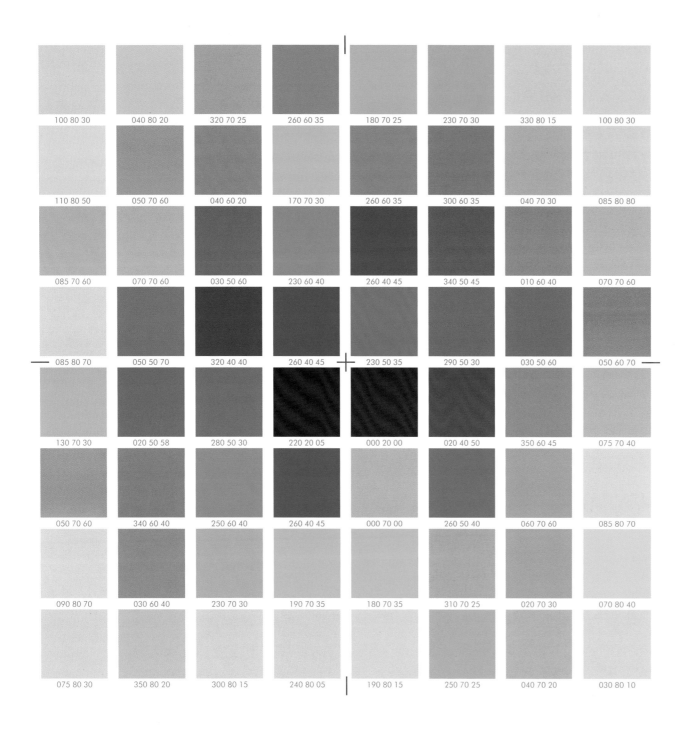

DISCERNING
谨慎挑剔

EXPENSIVE
贵气逼人

ENRICHING
TOP-NOTCH
DISCERNING
EXPENSIVE

丰富多彩
首屈一指
谨慎挑剔
贵气逼人

The interpretation of the four terms is predominantly about tonalities, which are positioned in the centre of the comprehensive colour spectrum. The cool shades and the warm ones balance each other out. The blue sequence is especially comprehensive, whereas the sedate, warm hues fade into the background.

When the expensive, the discerning, the top notch and the enriching vie for the favour of customers, then the colour values of clarity and unambiguousness are in demand. Shades that are tempered or restrained or too bright are pushed into the background. One also realises that aggressive, super-elegant glamour appeals to the senses are less visible than nuances that are fresher, activating, and that generate lucidity. Their dynamic is stronger than the features of other design attributes that draw on higher-status and elitist claims.

对于这四个方面的诠释主要体现于色调方面,这也是综合性色谱中的核心问题。冷色调和暖色调相协调。蓝色持续出现而不失稳重,暖色调渐渐消融于背景中。

贵气逼人、谨慎挑剔、首屈一指抑或是丰富多彩,哪种风格的色系会受客户青睐呢?此时色值较清晰的风格将脱颖而出。温和、素净或太亮的色调将被安排为背景色。人们也将意识到,鲜艳、活力迸发和清晰的色彩,较热烈奔放或超级优雅的色彩更吸睛,它们比其他声称更高档或更受精英喜爱的色彩更具活力。

TOP-NOTCH 首屈一指

Great expectations require top designs. Homeliness and absolute wellbeing with a high level of relaxation are preconditions for the fulfilment of this popular promise. Cool splendour is not at the top of the list, but rather a pleasant colourfulness and, frequently, more rounded forms than usual. Angular or even pointed surfaces and all such patterns, not to mention zigzag designs, should not be part of this sort of ambience. All in all, the design becomes more rounded, including spherical lamps and containers, but also other accessories and many types of furniture. Anything that one can cuddle up to and that feels cosy is in demand.

030 50 60

A pyramid is even more difficult to feel than a cube is. That which is round has the advantage of having a pleasant surface feeling, and is preferred. Once again, soft shapes and tactile qualities are winners: velour fabrics are just as en vogue as voluminous carpets and rounded armchairs, as well as bolide-like ceiling shapes, sofas and floor lamps.

高期望值需要顶尖的设计与之相配。如家一般的舒适自在、绝对的安乐和令人高度放松的酒店特质都是实现"首屈一指"这个承诺的先决条件。其首要任务不在于打造宏大的外观,而是营造宜人的色彩。

此外,通常还包括比平常更圆润的造型。有棱角或尖锐的表面以及此类的图案,更别提锯齿形状设计,不应是这种氛围的一部分。总而言之,设计,包括球形灯和容器,还有其他配饰和许多家具,应更圆润。任何可以让客人拥抱、给人舒适感的物什都是所需的。

030 50 60　040 70 30　100 80 30

比起立方体,角锥体给人更不舒服的感觉。在各种形状中,圆形在营造舒适感方面更有优势,也常被列为优选。案例中有着柔和形状和质感的物什再次成为赢家:丝绒织物、大地毯、圆形扶手椅都非常时尚,火流星般的吊顶造型、沙发和落地灯同样动人。

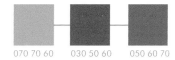

Not only the design must be top-notch, but also the quality of the materials and surfaces, and the functional value should be above average. It is best if tables and chairs do not convey the impression of fragility but of stability, if the desk does not wobble at the slightest push, and the clothes hangers do not fall apart when they have to support a slightly heavier load. A generous seating area and the comfortable conference table reveal a solid make and substance. Just like the well-known light switch with the large control panel.

不只设计必须一流,所选材料、表面质感和功能值也须高于平均水平。如若给人留下了桌椅稳定而不易损、即使被推桌子并不摇晃、衣架坚实的印象,这是最好的。而宽敞的座位区、舒适的会议桌和有着硕大控制面板的知名品牌照明开关同样体现了酒店的优秀品质。

| 030 50 60 | 070 70 60 | 085 80 80 | 050 60 70 |

010 60 25
030 50 60
040 70 30

HOTELS WITH REGENERATIVE FACTOR: GENTLE TO APPETISING

There are different ways to unwind from everyday routines. One option would be to go out into the countryside and help a friend who is a farmer to gather the harvest. Alternatively, one could stay at home, cut all forms of communication and read Goethe's theory of colours, and maybe also the conversations with his close friend Eckermann. A third option would be a quick trip around the world, paying a visit to the ten best opera houses.

An alternative that comes to mind is to get in a car, on a train or aeroplane and take oneself to a beautiful, cosy, comfortable, scenic place and indulge in the pleasures of being relaxed and lounging around. This should include all-around care, either more or less physical exercise, plenty of or little water, sauna and massages.

It would be beneficial to relearn how to stroll. Strolling is more than mere rambling. The stroller takes part in nature, observes the shop window displays, physiognomies of people, their attitudes, gestures and vanities. Strolling includes going to cafés, parks and quiet, secluded benches underneath climbing roses and dense foliage. For strolling, one needs a conversation partner as well as meeting points inside or outside the hotel.

As a matter of course, a beautiful lobby with soft settees and extremely comfortable deep armchairs is part of this. Dimmed light and an unrestricted view of arriving guests are very welcome, and so is a view of the hotel garden and the landscape in the background.

The employees looking after the guests at all levels of the hotel hierarchy are the most important ghosts, the true philosophers and well-informed attendants of the initially insecure guests, whose familiarisation with the new environment may sometimes be slightly complicated at first. They need tips and friendly advice on how their days can take a pleasant course even without duties.

If the hotel is equipped with a library, everything is a lot easier. Evening readings or discussion groups initiated by the hotel management work miracles. It is great if there is a philosopher is in the area who has reflected on "being" and is a clever narrator. One should hurry to engage him. Local colour is well received.

Organised events of this kind in the hotel library are gratefully accepted by experienced guests; having a night cap with one's bed close by makes things very easy.

生机盎然的酒店：
从体贴入微到秀色可餐

从日常事务中脱离出来的方式有多种。一种方式是走出城市来到乡村，帮助农民朋友收割庄稼。另外，人们可以待在家里，断绝与外界联系的一切形式，静静阅读歌德的色彩理论，或与歌德的亲密朋友爱克曼对话。第三种选择即是在世界各地短途旅行，参观十家最佳的歌剧院。

另外，也可通过汽车、火车或飞机，去往美丽而温馨、舒适而风景秀丽的地方，尽情享受放松和闲逛的乐趣。这应包括全方位的呵护、或多或少的体育锻炼、充足的或少许水，以及桑拿和按摩服务。

这将有利于人们重新学习如何去闲逛。闲逛不只是无目的地漫步，人们可以融入自然，欣赏商店的橱窗，观察人们的相貌、态度、举止和手势。可以闲逛至咖啡馆、公园或掩映在攀援的玫瑰和稠密的植物下的长椅，此时人们需要一个交谈的对象和酒店内或酒店外的见面地点。

当然，设有软长椅和舒适扶手椅的华丽大堂也是其中的一部分。暗淡的灯光，以及可以看到所有来客的毫无遮挡的视野深受欢迎。同样，酒店花园的美景和酒店后的景观可以一览无遗，这也广受好评。

照顾客人的各级酒店员工仿佛"幽灵"，他们非常重要，对刚刚到来还缺乏安全感的客人来说，他们是哲学家，是消息灵通的服务员。熟悉新环境一开始比较复杂，客人需要酒店员工的提示和友好建议（甚至是职务外的），以在酒店里度过美好的时光。

如果酒店配备了图书馆，一切将轻松许多。酒店发起的晚间阅读和讨论小组将产生奇迹。如果讨论成员中有反思"存在"的哲学家或好的解说员，人们更会抢着加入。此外，有地方色彩的话题一致得到好评。

这种在酒店图书馆组织的活动将会为经验丰富的客人欣然接受；此外，床头备一睡帽将使入住更为惬意。

104	gentle - regenerative - quiet - neat	体贴入微 – 生机盎然 – 静水深流 – 简洁淡雅
118	comfortable - rejuvenating - invigorating - stimulating	舒适惬意 – 焕发新生 – 万物生长 – 振奋人心
132	relaxing - soothing - serene - sensitive	轻松自在 – 轻柔舒缓 – 静谧安详 – 敏感细腻
148	healing - refreshing - vitalising - appetising	情感治愈 – 清凉酷爽 – 无限活力 – 秀色可餐

GENTLE

体贴入微

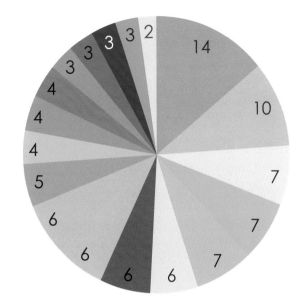

The colourful pie chart shows how to interpret "gentle dealing" with colour shades. "Gentle" is very rarely grey and never bright white.

多彩的饼形图显示了如何以色调诠释"体贴入微","体贴入微"风格的色调很少有灰色,从未有亮白色。

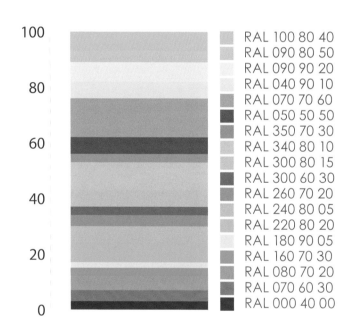

RAL 100 80 40
RAL 090 80 50
RAL 090 90 20
RAL 040 90 10
RAL 070 70 60
RAL 050 50 50
RAL 350 70 30
RAL 340 80 10
RAL 300 80 15
RAL 300 60 30
RAL 260 70 20
RAL 240 80 05
RAL 220 80 20
RAL 180 90 05
RAL 160 70 30
RAL 080 70 20
RAL 070 60 30
RAL 000 40 00

A wonderful colour scale is the result. It has a great design potential because it can operate effectively with the bright surfaces which look as if they have been dusted with powder.

美妙的色彩有着巨大的设计潜力,有看似施以脂粉而鲜亮的表面,易于使用。

No exciting shapes; they mostly appear flat and are occasionally striped.

没有令人兴奋的形状;它们大多扁平,偶尔有条纹出现。

GENTLE 体贴入微

A few hints of child-like-fairytale rosé and pink are just as important for the overall appearance as the spring-like shades of mimosa. Almost every colour field exhibits light, multi-toned interpretations of cloud pictures.

含羞草般、春意盎然的色调对整体外观的塑造非常重要，些许童话般的玫瑰色和粉红色也同样重要。几乎每个色域的图都诠释了多种色调、浅浅的云朵。

REGENERATIVE

生机盎然

Wonderfully colourful, appealing and narrative. Slightly disordered, but extremely attractive and thus effective as an ensemble.

色盘非常丰富多彩，有故事性，有吸引力，看起来稍显紊乱，但色彩的合奏极其美妙动人。

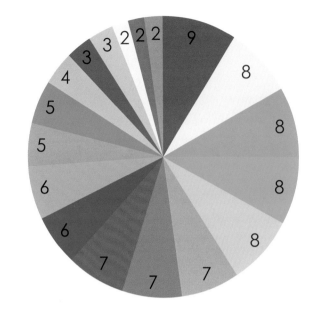

Individual segments of the bar chart can also be used effectively: for example, the green-golden, the orange-red and the blue-brown colour fields.

条形图各段均可有效使用，例如绿金色、橙色、红色、蓝色、褐色色域。

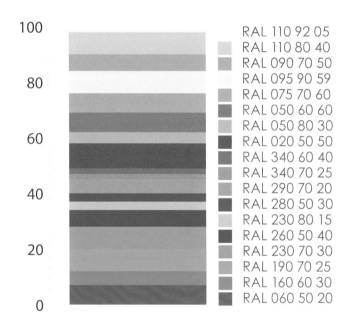

RAL 110 92 05
RAL 110 80 40
RAL 090 70 50
RAL 095 90 59
RAL 075 70 60
RAL 050 60 60
RAL 050 80 30
RAL 020 50 50
RAL 340 60 40
RAL 340 70 25
RAL 290 70 20
RAL 280 50 30
RAL 230 80 15
RAL 260 50 40
RAL 230 70 30
RAL 190 70 25
RAL 160 60 30
RAL 060 50 20

Diagonal, vertical and simple horizontal graphical signs make up the repertoire of shapes.

对角、竖直和简单的横向图标组成了全部形状。

REGENERATIVE　　生机盎然

It cannot be denied that the basic attitude is optimistic. The shades serve to improve that basic mood because they are clean, clear and superficial. They smell of flower garden and taste of fruit cocktail.

不可否认的是，它的基调是乐观的。这些色调有助于改善情绪，因为它们干净明确，且较浅，仿佛带来了花园中鲜花和什锦水果的芬芳。

QUIET

静水深流

In its weighting, the orchestra of tones consists of the components of air, water, winter and spring, silent serenity and immobility.

从各种色彩在色盘中所占比重可以看出，这首色调奏鸣曲中包含了空气、水、冬季和早春、静默无声和静止的意象。

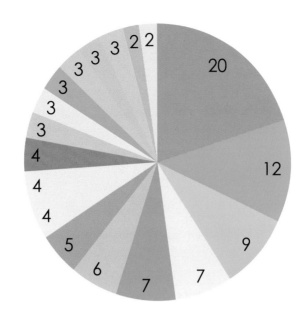

The message indicates that quiet cannot be accentuated. One cannot help but notice the consistent restrictions on the interpretation exerted here.

条形图所给信息表明宁静风格的色调不可过于突出。会令人不禁去注意的是，从"静水深流"色调中所理解出的都是束缚感。

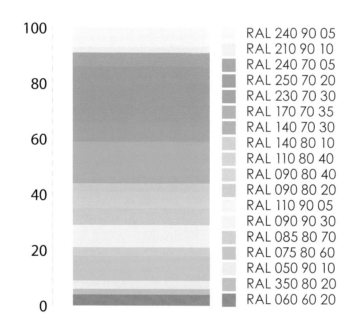

The signs indicate forms of simple contents like amorphous shapes, circles, and double bars.

这些符号表明其形状大多简单，如圆形、复纵线或者不规矩的形状。

QUIET   静水深流

White does not convey quiet. Quiet does not mean soundlessness but rather the desire for a distant murmur or gentle background music. The quiet before the storm is overpowering, the one afterwards is heavenly.

白色不能表示宁静。宁静并不意味着万籁俱寂，反而是对遥远的轻吟或轻柔背景音乐的渴望。风暴来临前的宁静无法抵挡，风暴过后的宁静则妙不可言。

NEAT

简洁淡雅

The colour circle reveals above all freshness and conspicuously pale shades. Bombastic effects are not part of this theme.

色盘上鲜亮和浅浅的色调最为突出。夸张的效果并非该主题的一部分。

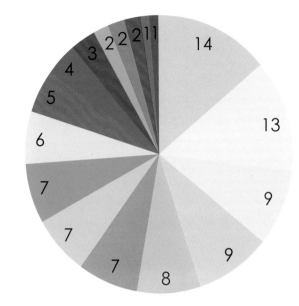

A lightweight sequence with colour shades predominately reminiscent of a sea breeze and the taste of salt. It has the capacity to calm and create a restful ambience.

色调由轻及重排列,让人联想到海风和盐的味道。它具有让人舒缓放松的作用。

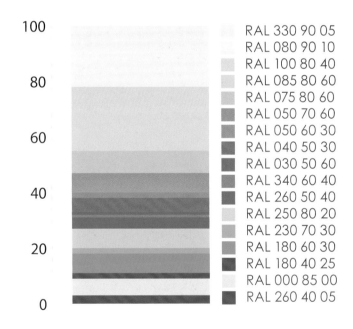

RAL 330 90 05
RAL 080 90 10
RAL 100 80 40
RAL 085 80 60
RAL 075 80 60
RAL 050 70 60
RAL 050 60 30
RAL 040 50 30
RAL 030 50 60
RAL 340 60 40
RAL 260 50 40
RAL 250 80 20
RAL 230 70 30
RAL 180 60 30
RAL 180 40 25
RAL 000 85 00
RAL 260 40 05

Drawing symbols are dispensable for the majority of interpreters. Only a lonely zigzag movement stands out.

这些图形符号对大多数的诠释者来说可有可无,只有锯齿波形运动走向引人注目。

NEAT   简洁淡雅

The majority of study participants decided in favour of an immaculate, tranquil colourfulness with a cosmetic quality. It is delicately powdered, as if applied with a powder puff, and it reveals delicate details.

大多数研究参与者更偏爱洁净的色彩。它们宁静且具有装饰性，仿佛扑过粉、化了妆的佳人，有着精致的细节。

GENTLE 体贴入微

REGENERATIVE 生机盎然

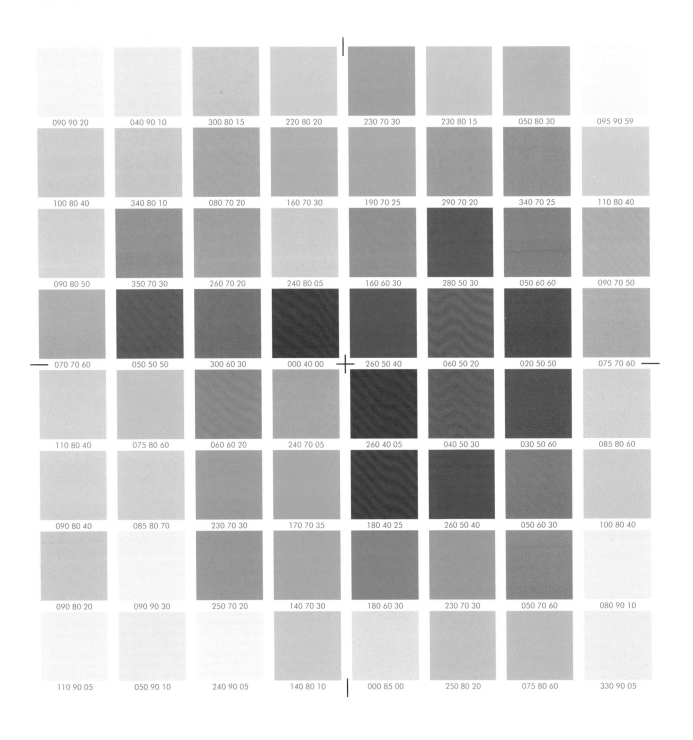

QUIET 静水深流

NEAT 简洁淡雅

GENTLE
REGENERATIVE
QUIET
NEAT

体贴入微
生机盎然
静水深流
简洁淡雅

The music of the colours is composed of cream and sorbet shades. A restorative, therapeutic background is clearly discernible. Colours composed with such intentions can be helpful and offer welcome support for the healthy guest in search of successful relaxation and enjoyable peace and quiet. Even the most delicate limestone hues are frequently furnished with a sufficient quantity of colour particles to create a pleasant, Please also take a look at the top and the bottom row on the page on the left.

Not one, two or three shades should define the range. Only a tonal range made up of up to nine colours, with a variety emphases, of course, conveys to the beholder the characteristic impression of that which is "gentle" or "regenerative". One should not hesitate to use the depicted colour spectrum for the overall impression, and here it is worth reiterating that one to two shades should determine the impression; the others act as accompaniments.

这首色彩奏鸣曲由奶油色调和冰糕色调组成。很明显，它们是疗养、放松的背景。出于此意图，它们有助于客人寻求放松、平和、安宁。即使是最微妙的灰白色调也常配以众多彩色颗粒，以营造愉悦而富有创造力的氛围。

此风格色彩有不止一种、两种或三种色调。虽只由九种颜色组成，却各有重点，传达给观者的印象特点是"温和的"或"有再造能力的"。在营造整体印象时，人们不应吝惜使用此色谱。值得强调的是，应以其中一到两个色调奠定基础，其他只作为辅助。

NEAT

简洁淡雅

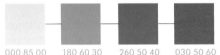

000 85 00   180 60 30   260 50 40   030 50 60

An ambience that radiates and promotes care has the charm of neutrality with a touch of emotion; naturally, this is at the expense of a more unrestrained individualisation of the interior design. The pages that follow reveal how this can be achieved in an entertaining way. Tinted sorbet and cream colours prepare the living environment: the walls are prepared with animated wool-linen or cream-toned colours. Different light sources illuminate the room in sections, as well as the larger spaces like the lobby, restaurants and the bar. Each corner requires its own identity. The view of the outside reveals the strongest contrast.

The format of the pictures should be as large as possible. The minimum size, regardless of the dimensions of the rooms themselves, should rarely be less than 40 x 60 cm, because even in small rooms, large pictures do not reduce but enlarge the space.

The hues used show sufficient shadings because several wood, wool and chalk hues frequently supply enough colourfulness to create a lively, albeit limited, atmospheric environment. This is supplemented by the more shadowy and illuminated sides that create invigorating visual events. Horizontal wall elements enlarge the room and add calming stability. Vertical directions, by contrast, are more stimulating and reduce the dimensions.

040 50 30   085 80 60   080 90 10   050 70 60

080 90 10   050 60 30   040 50 30

　　流露出关怀的氛围有着淡淡的情感魅力，当然，这要求室内设计不可受限太多，而应更奔放、个性化。接下来的内容将展示它以何种愉快的方式实现。设计以奶油色调和冰糕色调营造居住环境：墙壁均采用了活泼的羊毛、亚麻或奶油色调的颜色。不同光源照亮了房间的不同部分，以及大堂、餐厅、酒吧等较大空间。每个角落都要有自己的特点。外界的风景更能凸显这种强烈的反差。

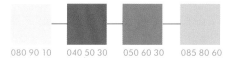
080 90 10   040 50 30   050 60 30   085 80 60

040 50 30   050 60 30   080 90 10

080 90 10   330 90 05

Small and large accessories, ranging from cushions, lamps and a number of candles to quilts, trays with vases, glasses, ceramics and other everyday objects, make up the coloured and decorative part of the interior. Plain jacquard fabrics with raised and velour effects, in this case without large patterns, as well as comfortable seating furniture are other elements that contribute to wellbeing.

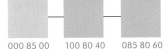
000 85 00   100 80 40   085 80 60

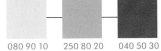
080 90 10   250 80 20   040 50 30

壁画应该尽可能大。不管房间本身大小如何，它最小的尺寸都很少小于40厘米x60厘米，因为即使在小房间内，大大的壁画在视觉上不会缩小而是放大空间。

所使用的色调明暗对比恰到好处，木材的色彩、羊毛色和粉笔白色足以营造活泼而不至于太过跳跃、大气的环境。它们在光与影的辅助下，营造出清爽宜人的视觉印象。墙上横向的元素在视觉上放大了空间。竖直方向的元素则使房间看上去略失稳重且缩小了空间。

大大小小的配饰，从靠垫、台灯、蜡烛到被子、托盘、花瓶、玻璃、陶瓷以及其他日常用品，构成了多彩的室内空间装饰成分。此外，还有素色、无大面积图案的丝绒提花面料、舒适的座椅和家具等提升了舒适度。

080 90 10

050 60 30

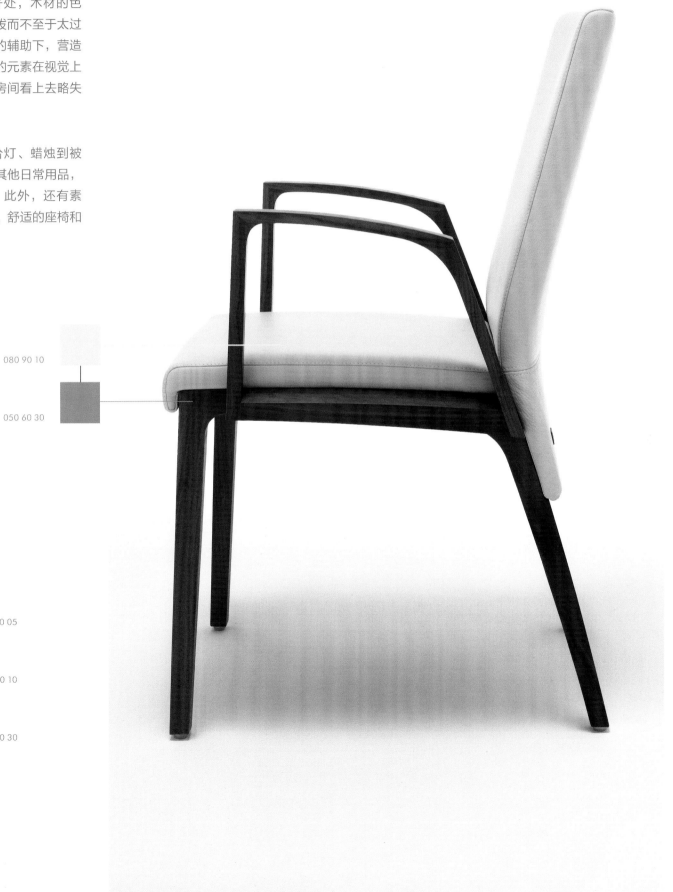

330 90 05

080 90 10

050 60 30

COMFORTABLE

舒适惬意

Except for the occasional bright hues, the pie chart displays relatively heavy colours. On the whole, it is reminiscent of soft cushions, woollen blankets, wingback armchairs or coarse-knit, patterned winter jumpers.

亮色调偶尔才出现，饼形图展示了较重的色彩。总体上，它很容易让人联想到软靠垫、毛毯、后腰扶手椅或粗针织物，以及冬天里有图案的毛衣。

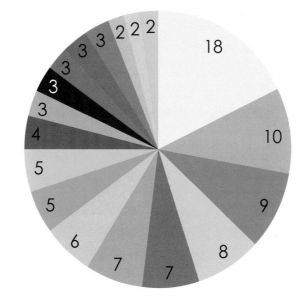

The range features the rural charm of a cottage garden in October. It radiates cosiness and a certain degree of determination.

该系列散发着十月里乡间花园的乡村魅力。它既给人舒适感，一定程度上也显得很利落。

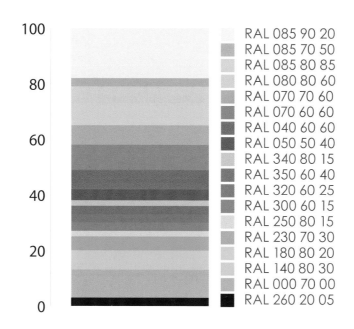

Couches and the shapes of rocking chairs and sofa cushions indicate the characteristics inviting us to relax.

沙发、摇椅和沙发靠垫的造型看起来很友好，好像在邀请客人落座放松一下。

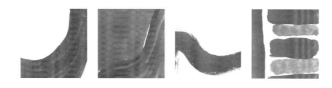

COMFORTABLE 　　　　　　　　　　　舒适惬意

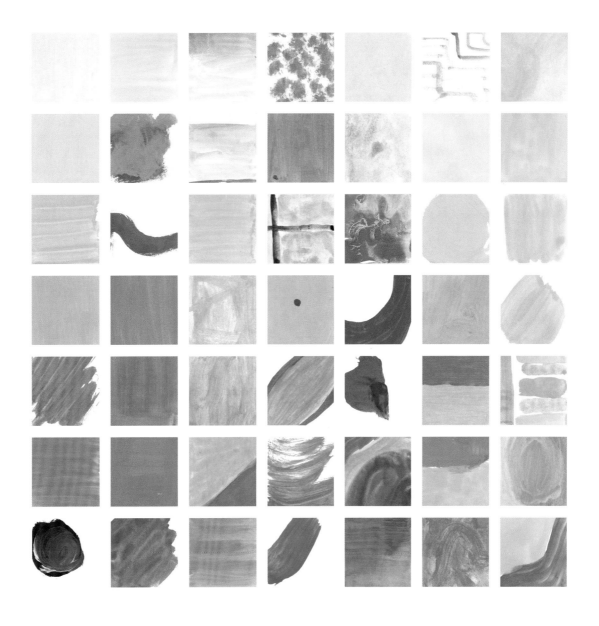

The colour scheme of restfulness and comfort rest and cosiness is free of bright and lively shades. The overall impression is of movement at a steady, even a sedate pace, and comfortable satiety. The colours have overtones of potatoes, earth and the warmth of autumn.

"舒适惬意"风格的配色方案不会有太鲜亮、活泼的色调。总体给人留下淡雅、稳重的印象，此外还有舒适和满足感。这种风格的色彩使人联想到土豆、泥土和秋日的温暖。

/ Chapter 2

REJUVENATING

焕发新生

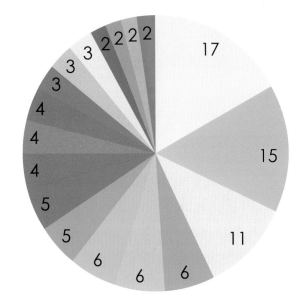

The colour circle clearly illustrates that both warm and understatedly fresh hues belong to the repertoire of regeneration. Black and white hardly ever occur.

色盘清楚地说明了温暖、清新朴素的色调都属于复兴风格。黑与白色极少出现。

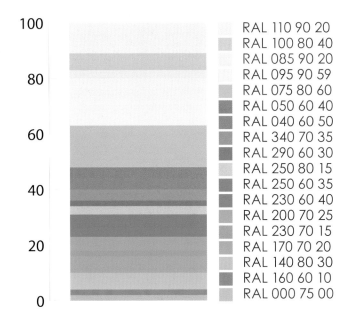

The chart shows a colour scheme that focuses on the centre. The shades are strongly mixed to some extent so that unrest and sudden excitement are virtually non-existent.

该系列展示了一种突出中心的配色方案。色调经过强烈混合，在一定程度上使得动荡感和突然的兴奋感几乎不存在。

The design repertoire is limited. Except for a few hints reminiscent of plants the sequence remains calm and free of movement.

图案中所运用的设计技巧十分有限，除了一些形状让人想起植物，整个序列平稳、缺乏动感。

REJUVENATING　　焕发新生

Summery, almost vernal green, yellow-orange and bright blue shades enliven us. The colour message of revival points to agitation with dynamic limits. The process of regeneration needs hues that are anything but brilliant.

满载夏日甚至春天气息的绿色、黄橙色和亮蓝色能让我们兴高采烈起来。复兴风格的色彩信息易传递一定程度的躁动，"新生"过程需要的是绚丽的色彩。

INVIGORATING

万物生长

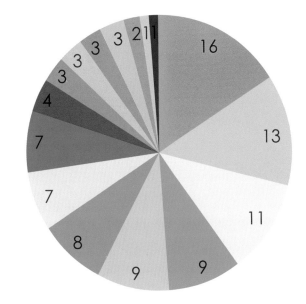

The interpreters have imitated what nature offers. The liveliness of the artistic paint application can be seen in the colour circle.

仿佛大自然给予我们的色彩，色盘中可看到艺术涂料的应用产生的生机勃勃的效果。

The result is a harmonious, summery coloured. It is as fresh as water, as warming as the sun and as encouraging as burgeoning greenery.

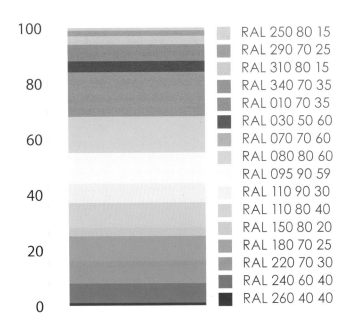

RAL 250 80 15
RAL 290 70 25
RAL 310 80 15
RAL 340 70 35
RAL 010 70 35
RAL 030 50 60
RAL 070 70 60
RAL 080 80 60
RAL 095 90 59
RAL 110 90 30
RAL 110 80 40
RAL 150 80 20
RAL 180 70 25
RAL 220 70 30
RAL 240 60 40
RAL 260 40 40

成果便是和谐的、洋溢着夏日气息的色彩。它如水般新鲜，如太阳般温暖，如生机勃勃的草木般鼓舞人心。

Round and linear elements occur as frequently as dot-shaped structures and light runs.

圆形、线型元素与点状结构、顺串图形的出现同样频繁。

INVIGORATING　　万物生长

The overall image appears to be inspired by plants. The complete picture seems strongly narrative and cheerful, but serene and sometimes airy. Reddish and bluish basic shades take over most of the interpretation.

整体形象似乎是受了植物的启发。完整图片看起来更有故事性、更令人愉快,但它们宁静、通透。大部分以偏红、偏蓝色调来诠释。

STIMULATING

Cool nuances are in the majority. The bright pink and rosé hues participate in lively manner in the interplay of the airy presentation.

振奋人心

大多数为冷色调。在这个清晰的展示中，明亮的粉红色调和玫瑰色调以跳跃的方式与其他色调相互作用。

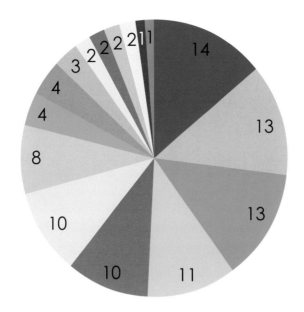

Thanks to the predominance of blue, the colour sequence appears transparent and ordered. The clarity of the colour sequence is only interrupted by a few hints of yellow and green.

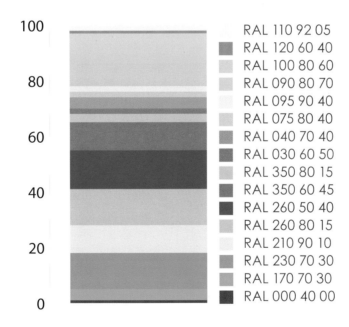

RAL 110 92 05
RAL 120 60 40
RAL 100 80 60
RAL 090 80 70
RAL 095 90 40
RAL 075 80 40
RAL 040 70 40
RAL 030 60 50
RAL 350 80 15
RAL 350 60 45
RAL 260 50 40
RAL 260 80 15
RAL 210 90 10
RAL 230 70 30
RAL 170 70 30
RAL 000 40 00

由于蓝色居多，色序看起来透明而整齐。清晰的顺序仅因少许黄色和绿色中断。

Time and again, the wings of small and large animals flying in the air are represented. In this way, the study participants created miniature paintings.

大大小小的动物一次又一次在空中飞行，它们的翅膀屡屡出现。通过这种方式，研究参与者创作了许多袖珍画。

STIMULATING　　　振奋人心

The blue variations occur frequently and conjure up the impression of atmospheric lightness and radiant brightness. The other shades accompany the colours of the sky so that everything seems to be hovering.

深浅不同的蓝色频繁出现，频繁变化，使人想起了大气的轻盈和亮度。其他色调衬托着天空的颜色，一切看起来似乎都在天空中盘旋。

COMFORTABLE
舒适惬意

REJUVENATING
焕发新生

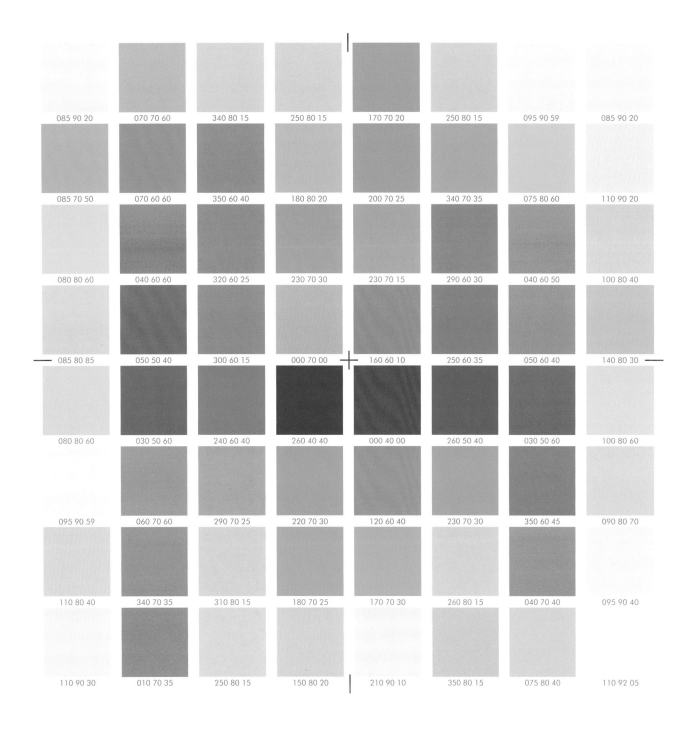

INVIGORATING
万物生长

STIMULATING
振奋人心

COMFORTABLE
REJUVENATING
INVIGORATING
STIMULATING

舒适惬意
焕发新生
万物生长
振奋人心

As if inspired by spring awakenings, these shades illuminate floral pictures of colour, which unfold the awakening of nature and its invigorating effects before our eyes. All they lack is the saturation of summer. The colouristic impression is entirely free of murkiness. Only one or two light shades of grey have found their way into this colour range.

The colours undoubtedly express an optimistic mood. They convey a gentle thrust of merry serenity and well-tempered sensual pleasures. The number of light botanical plant colours outweighs the air and water hues, which also occur in large numbers. One can live very happily among this colour sequence, and also survive any cool days that come along. They have great therapeutic qualities, contrary to the shades on the opposite page that contain grey, black and violet, whose effectiveness tends towards more melancholic moods.

仿佛受到万物复苏的春天景象的启发，这些色调点亮了这幅由色彩组成的花海图。展现了大自然的美，具有振奋人心的效果。它们所缺乏的是夏天的饱和度。这些色彩印象完全与阴暗无关。在这种风格色彩中，只有一两种浅灰色调。

毫无疑问，这些色彩体现了乐观的心态，传达出欢快、爽朗、愉悦之感。虽然天空与水流的色彩大量出现，但与之相比，浅植物色彩出现的更多。身处此种色彩序列中，人们可以非常欢乐地生活，度过任何严酷的日子。与较忧郁的灰色、黑色和紫色色调相比，它们有着良好的治愈特性。

REJUVENATING 焕发新生

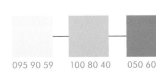
095 90 59   100 80 40   050 60 4

The therapeutic purpose of staying in a hotel is becoming increasingly essential. "Just unwinding for a few days" suits the imagery in the same way as "eight days of wellness, fitness and enjoyment" or "a fortnight of chilling out with daily visits to the hammam, a diet and the perfect coach". What was once prescribed as "a health cure in the countryside" or "a convalescent holiday in a clinic", and took place in a barrack-like atmosphere, is today offered by the most beautiful hotels with various forms of service.

110 90 20   085 90 20   095 90 59

In the "Release From Daily Routines" category, hotels enjoy an excellent reputation. Walls are white only where it is necessary, for example for hygiene-related reasons. The facility variations generate pleasant feelings of comfort, security and freshness, as well as health, revitalisation and activation processes.

Especially colours, but also shapes, structures and patterns, can contribute significantly to making us feel better in a suitably designed environment. Colours, as well as light and warmth, influence our physical and psychological quality of life. We prefer vernal shades while engaging in recreation to grey, black, blue and brown. We perceive warming colours as more congenial and helpful than very cool hues. Shades of green, yellow and orange have refreshing and exhilarating effects.

出于疗养目的而住酒店的趋势正在上升。"放松几天"、"轻松、健身和享受一周"、"去放松两星期，泡泡澡，减减肥，跟着教练健身"逐渐和酒店联系起来。曾经一度被视为"在乡村做疗养"、"在诊所做康复训练"的正事，现在都可在提供各种服务的漂亮酒店内完成。

酒店在"释放日常压力"的休闲娱乐界享有良好的声誉。只有在必要时，例如卫生方面的原因，墙壁才选用白色。设施的变化带来了舒适感、安全感和新鲜度，有益于消除疲惫、恢复体力和振奋精神。

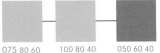
075 80 60   100 80 40   050 60 40

075 80 60   085 90 20   050 60 40

200 70 25   075 80 60   085 90 20

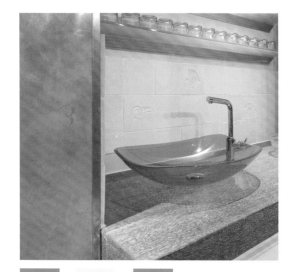

040 60 50   085 90 20   050 60 40

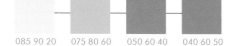

085 90 20   075 80 60   050 60 40   040 60 50

We definitely like round shapes, or shapes one can cuddle up to, better than triangles and pyramids or squares and cubes. Sharp edges and acute angles do not inspire recreation. We are especially fond of longitudinal stripes because they point us in the right direction. We like hideaways in particular because they screen us off, partly conceal us and give us shelter but simultaneously allow us to observe everything from a safe distance with wonderful curiosity.

形状、结构和图案，尤其是颜色，能影响我们在设计恰当的环境中的舒适感。色彩、光线和温暖度，影响着我们的生活质量。我们更喜欢春天般的色调，同时以休闲的灰色、黑色、蓝色和棕色搭配。我们认为暖色调比冷色调更适宜、更有益。绿、黄、橙色调具有令人清爽、令人振奋的功效。

我们十分喜爱圆形或其他可以抱住的形状，而非三角形、角锥体、方形或立方体。尖锐的边缘和角度对消遣无益。纵向条纹也备受我们喜爱，因为它为我们指明了方向。此外，有个藏身之处最好不过，它可遮挡我们，将我们部分隐藏起来，提供了遮蔽；同时，它允许我们在安全的距离内兴致勃勃地观察一切。

050 60 40

085 90 20

100 80 40

085 90 20

100 80 40

050 60 40

075 80 60

RELAXING

轻松自在

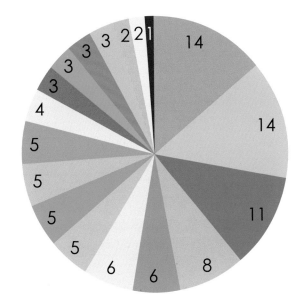

According to the colour content and intensity, relaxation is not only a passive process, but, as the study participants observed, it also has features which demand active participation.

此风格的色调在色彩内容和强度上不仅仅是被动的，研究参与者观察到，它也有需要积极参与的色彩内容。

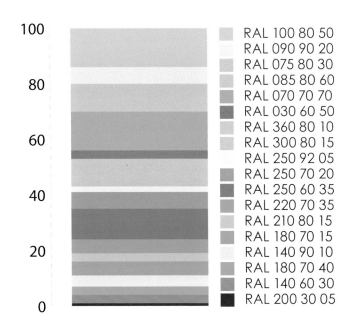

This sequence contains two essential scales: a green, yellow, orange section and a rosé, light blue, turquoise-coloured one.

该序列含有两个基本范围：其一为绿色、黄色、橙色部分，其二包含了玫瑰色、淡蓝色和绿松石色。

Signs representing the water, the sun and the air serve as signals of the perfectly nap, or in any case stress-free relaxation.

这些符号代指水流、太阳和空气，同时也是代表舒适小憩或其他任何毫无压力的放松状态的符号。

RELAXING 轻松自在

Shades that look like finest Italian ice cream or as if they come from the colour repertoire of an imaginative barkeeper in St. Tropez, Marbella or Brighton. Shades that can be experienced by the senses of smell or taste as well as visually.

它们看起来像意大利最好的冰淇淋的色彩，或来自于圣特罗佩、马贝拉或布莱顿富有想象力的酒保的配色。这些色彩不仅可以借助视觉来感知，还可凭借嗅觉和味觉来体验。

SOOTHING

轻柔舒缓

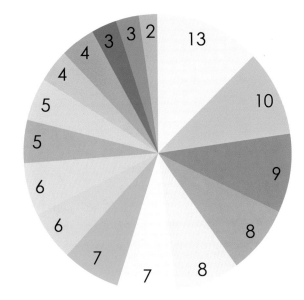

The pure shades of the sun, like red and vibrant orange, do not exist here. All colours are well tempered: not too hot and not too cold, but rather bright to pastel and unclouded.

这里不存在红色和活力橙色等纯色调。所有的色彩都比较温和，不会太热烈，也不会太冷酷，但鲜亮、柔和而明朗。

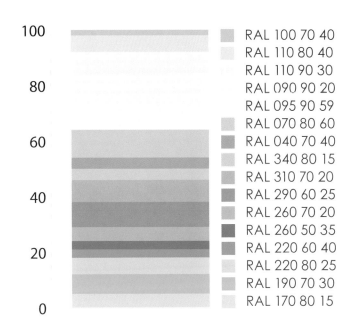

Pale shades suggest a creamy gentleness. Easy-care, beneficial and ageless.

RAL 100 70 40		
RAL 110 80 40		
RAL 110 90 30		
RAL 090 90 20		
RAL 095 90 59		
RAL 070 80 60		
RAL 040 70 40		
RAL 340 80 15		
RAL 310 70 20		
RAL 290 60 25		
RAL 260 70 20		
RAL 260 50 35		
RAL 220 60 40		
RAL 220 80 25		
RAL 190 70 30		
RAL 170 80 15		

浅浅的色调使人想起温和的奶油色，易打理、效果好且经久不衰。

Gentle waves, swirling clouds and blurred surfaces make up the graphical playing field.

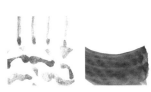

温柔的海浪、卷云和模糊不清的表面构成了生动的运动场。

SOOTHING 轻柔舒缓

The active mechanisms seem to be both female as well as male in nature. This is indicated by the blue shades on the one hand, and the feminine, mystical violet hues on the other.

这些色调搭配得非常灵活，有的阳刚，有的娇柔。一方面，这从蓝色色调可以看出来，另一方面，从神秘而女性化的紫罗兰色调也可得出此结论。

SERENE

静谧安详

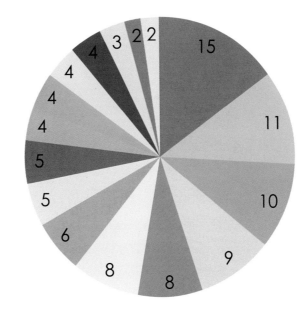

Almost too much gentleness and too much beauty for the moment. These shades are better seen as a temporary loan rather than a permanent possession.

此刻太过柔和、太过美丽。比起长期使用来，这些色调只作临时色彩更好。

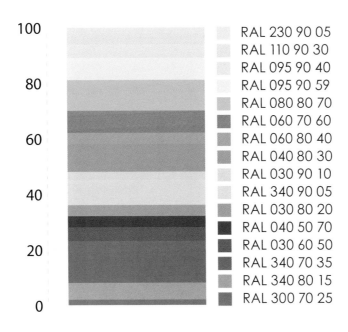

The colour sequence can be a good strategic model to document a paradigm shift from pale functionality to pure introspection.

该色序可作为很好的战略模板，成为如何从苍白无力转向打动人心的良好示范。

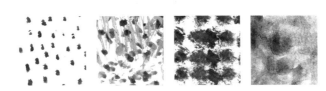

Small dots, wisps of clouds and tiny strokes guarantee this demonstration of the lovely-cute in all of us.

小圆点、缕缕云烟以及细致的笔触，充分体现了隐藏在所有人身上的可爱之处。

SERENE   静谧安详

The shades for special occasions. If we look at them, memories are brought back that range from the ridiculous to the sublime. Most eyes become soft and misty, at least for a few moments.

这些色调可用于特殊的场合。看着它们，我们会想起过去或荒诞或庄严的时刻，大多数人至少会眼眶湿润一会儿。

SENSITIVE

The shades are certainly subtle. Even the dark grey fits in well with the predominant sorbet-coloured lightness.

敏感细腻

这些色调自然是微妙的。即使是暗灰色也与主要的浅冰糕色配合无间。

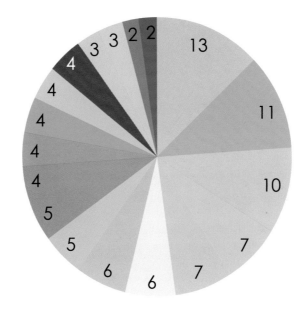

The colour sequence can be easily transferred to interior products: from the wall to the floor, from the windows to the upholstery fabrics and to the cushions, covers, and lampshades.

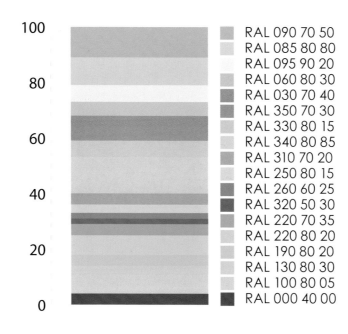

RAL 090 70 50
RAL 085 80 80
RAL 095 90 20
RAL 060 80 30
RAL 030 70 40
RAL 350 70 30
RAL 330 80 15
RAL 340 80 85
RAL 310 70 20
RAL 250 80 15
RAL 260 60 25
RAL 320 50 30
RAL 220 70 35
RAL 220 80 20
RAL 190 80 20
RAL 130 80 30
RAL 100 80 05
RAL 000 40 00

此色彩顺序易在室内出现：从墙壁到地板，从窗户到装饰面料和靠垫，从盖子到灯罩。

We recognise virtually nothing but graceful lines, small dabs and hints of structures.

除了优美的线条、小块和些许不太成形的结构，我们无法辨认出其他。

SENSITIVE   敏感细腻

Very bright shades have an affinity with cotton wool, christening robes and engagement presents. Many an elegant parlour dating from the turn of the 20th century flirts with ladylike grandeur like this.

非常明亮的色调与棉绒、洗礼长袍和订婚礼物有着密切的联系。许多19世纪末20世纪初的客厅便有着这种大家闺秀般的优雅。

RELAXING
轻松自在

SOOTHING
轻柔舒缓

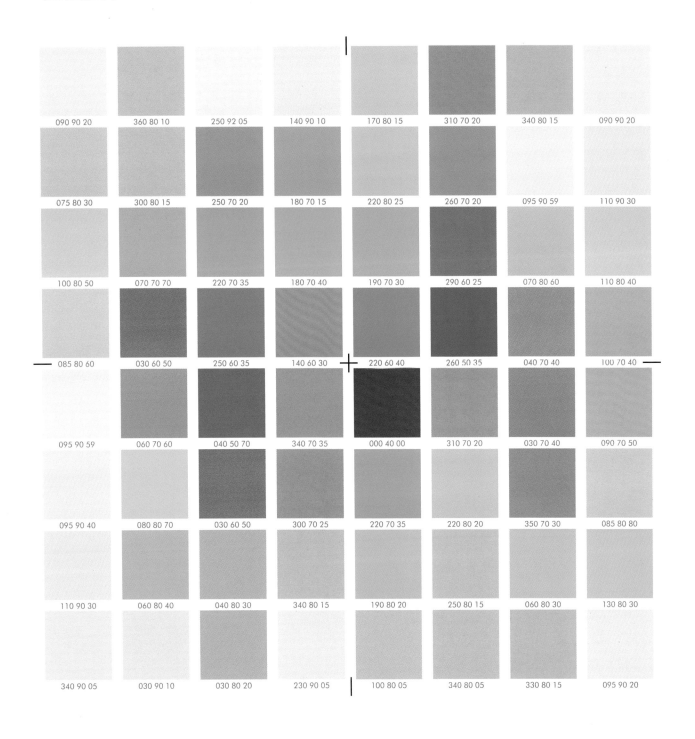

SERENE
静谧安详

SENSITIVE
敏感细腻

RELAXING
SOOTHING
SERENE
SENSITIVE

轻松自在
轻柔舒缓
静谧安详
敏感细腻

Wonderful interiors can be designed with just the four outside rows of the colour field that comprises sixty-four individual shades. Owing to their emphasis on harmony alone, they will create an enchanting ambience. One can see that the proportion of green is significant because delicate green brings relaxation and additionally provides us with the chlorophyll essential for survival. However, the light shades of grey also have a sophisticated, delicate and truly elegant effect.

The colours of the four themes feature very little brown. Their sensitive characteristics form a network of qualitative features with only little variance. Warm nuances prevail only for the term "gentle". The highest proportion of blue can be found in the interpretation of "soothing". The four colour overviews are certainly suitable for such ambiances in the form of single and overall spectrums. All shades feature a high level of purity and thus a significant degree of congeniality, as is shown by our colour research. Designing with these shades means offering the guest added value in terms of a personal sense of wellbeing.

想要达到最佳的室内效果，可以从这64个色块中选择边界线上的颜色。它们都很和谐，能够营造迷人的氛围。绿色占据很大比例，因为清新的绿色使人放松，而且带来生机与活力。同时，淡灰色也能营造精致而优雅的效果。

这四种风格仅使用极少量的棕色。风格的敏感性决定了它们相对固定的用色，变数极小。温暖的色彩是"体贴入微"风格的专属。"轻柔舒缓"风格则最青睐蓝色。根据我们的研究，这四个色块表中的颜色，无论单色还是混色，都与各自的风格相对应，承袭了浑然一体的特性。将这些颜色运用到设计中，可以为顾客带来真切的幸福感。

RELAXING　　　　　　　　　　　　　　　　　　轻松自在

Softness, lightness and the ability to relax are palpable features, as are the emotional characteristics of deceleration, contemplation and reduction. The design contents for the creation of a really relaxing atmosphere aim at bright, slightly clouded sorbet and pastel hues, at lilac, beech green, viola, agate green, water blue, dusky pink, nacre, lavender, creamy white and rose white.

090 90 20　　300 80 15　　075 80 30

250 92 05　　090 90 20　　100 80 50

The psychological profile of relaxation starts with the readiness to decelerate one's pace of life, to abandon curiosity, jealousy and envy. It is necessary to be in a wholesome spatial and scenic environment, not to get annoyed with oneself, and finally to do what one has always wanted to do: nothing, in other words indulging in dolce far niente, which literally means "sweet idleness".

Rooms and furnishings, colours and shapes all point to their ergonomic, relaxing and calming functions. Leaving all one's cares behind has to be learnt. This is done even more quickly and better if the messages are clear: the person looking for relaxation certainly needs no red cascades of colour or black-and-white appeals to the senses. Old Rolling Stones aficionados should be sensible and instead listen to Lana Del Rey. Light, imaginative, beguiling literature is to be recommended, whereas exciting crime novels are as poisonous as the current stock exchange report or the latest football results of one's favourite club.

柔软、轻盈、使人放松，是这一风格的明显特征，同时引导人们放慢节奏、静思冥想。设计旨在营造能够真正放松身心的氛围，选择明快柔和而略显朦胧的色彩，比如丁香紫、山毛榉绿、罗兰紫、玛瑙绿、水蓝、暗粉、珍珠白、熏衣紫、乳白和玫瑰白。

心理上的放松，要从放慢生活节奏开始，同时抛却好奇、羡慕和嫉妒的心态。必须在清新的空间和优美的氛围中，自我释然，最后只做一件事：放空，什么也不做，即所谓的"闲散享受"。

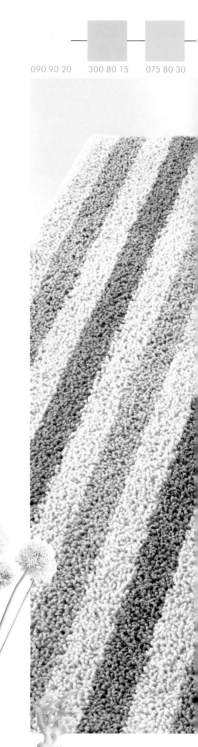

300 80 15   090 90 20

180 70 15   100 80 50   090 90 20

300 80 15   075 80 30   090 90 20

085 80 60   075 80 30

180 70 40   090 90 20   140 90 10

090 90 20

100 80 50

100 80 50

140 60 30

140 90 10

070 70 70

075 80 30

　　房间配饰的用色和形状都符合人体工程学原理，舒适无比。很重要的一点是，需要抛开所有的顾虑，只专注于寻找令人放松的元素，比如大面积的红色或正统的黑、白色就不适用。如果顾客是滚石乐队的歌迷，不妨换个心情听听拉娜·德雷。关于读物，轻松休闲、引人遐想的作品是上佳之选，而惊险刺激的犯罪小说最好放弃，因为这类小说和最新的股市报道、所钟爱俱乐部的足球比赛结果一样，足以搅乱原本宁静的思绪。

Being relaxed penetrates all one's senses. It comes all the more naturally if the conditions are geared towards tranquility. Noises, acoustic features, surface appearances and one's entire visual perception, which feeds more than ninety percent of consciousness and the subconscious, are among the means responsible for the sensory quality of beneficial spaces. The designer increasingly becomes a psychologist and philosopher, responsible for improving people's living conditions. Each colour, each pattern, each light source, each decision as to whether to use a patterned or a plain coloured fabric, a finely grained wood or smoothly varnished surface; everything must be well coordinated so that it pleases us.

075 80 30   100 80 50

保持放松，意味着所有感官都松懈下来。如果空间足够宁静，身心便能自然而然地得到放松。空间的噪声、声学特征、表面装饰以至所有的视觉感知，90%以上会反馈到一个人的意识或潜意识之中，所以这些元素成为空间感官品质的重要保证。设计师越来越多地充当心理师或思想家，需要揣度顾客的精神需求。挑选每一种颜色、图案、灯光，选择图案织物还是素色织物、细粒度木材还是光滑漆木，都需要设计师做出最佳的选择，以使所有的空间配饰都各得其所，最终给予顾客轻松之感。

090 90 20
140 90 10
300 80 15
075 80 30
100 80 50

075 80 30   300 80 15

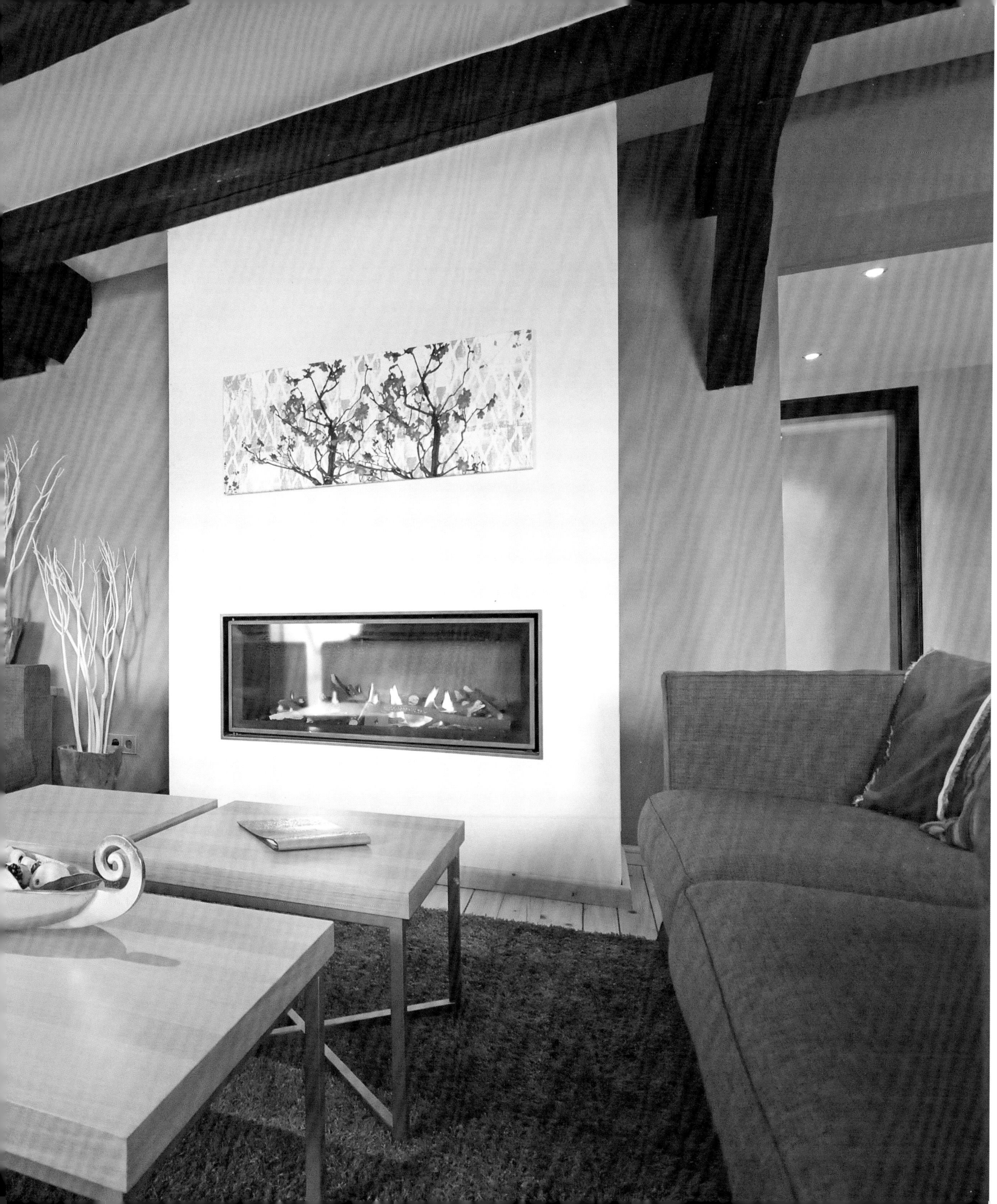

HEALING

情感治愈

The predominantly cool shades are accompanied and accentuated by buoyant hues of purple and scarlet.

清凉的蓝色与绿色成为主角，搭配活泼轻快的紫色与猩红色。

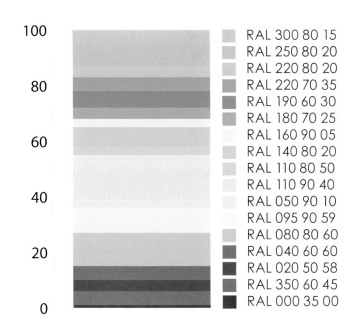

RAL 300 80 15
RAL 250 80 20
RAL 220 80 20
RAL 220 70 35
RAL 190 60 30
RAL 180 70 25
RAL 160 90 05
RAL 140 80 20
RAL 110 80 50
RAL 110 90 40
RAL 050 90 10
RAL 095 90 59
RAL 080 80 60
RAL 040 60 60
RAL 020 50 58
RAL 350 60 45
RAL 000 35 00

The colour stripes have a creative power. They emphasise the positive fact of convalescence and that the vitalisation process has been activated.

这些色条充满创造力，给人创伤康复、重获活力的积极暗示。

Signs from the health industry, which are caricatured with concentric features.

这些符号具有同心同轴的特征，是健康产业普遍采用的符号标识。

HEALING   情感治愈

The healing colour field shows a range of activating shades: blue, plenty of green, yellow, and a few hints of red display an amazingly lively reality. The process itself seems to be of central importance.

治愈系的色彩以蓝、绿、黄色为主,还有少量红色点缀其间,增添活力。对这些颜色在空间中的运用需用心把握。

REFRESHING

清凉酷爽

Not a single deep red pigment can be discerned. The green clover, grass and green apples are ultra-fresh. The soft sky blue is slightly warmer.

图中找不到一丝半点的深红色。三叶草绿、青苹果绿都无比清新。天空蓝则披上了一层温暖的色调。

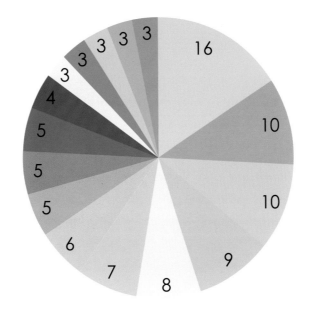

Colours which we like to experience when walking in the dunes on a sandy beach by the sea, because they do us good.

这些颜色让我们联想到漫步美丽海滩，沙区起起伏伏，如此清新怡人。

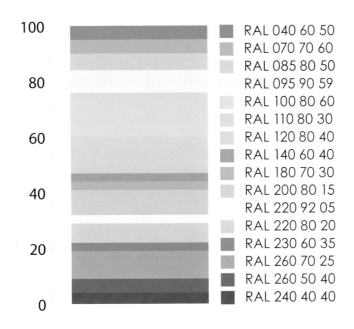

RAL 040 60 50
RAL 070 70 60
RAL 085 80 50
RAL 095 90 59
RAL 100 80 60
RAL 110 80 30
RAL 120 80 40
RAL 140 60 40
RAL 180 70 30
RAL 200 80 15
RAL 220 92 05
RAL 220 80 20
RAL 230 60 35
RAL 260 70 25
RAL 260 50 40
RAL 240 40 40

Goose pimples or snow flakes, in any case signs of a cool climate. If things are snappy, temperatures become arctic.

图案似鸡皮疙瘩，又似冰晶雪花，提示着酷爽气候的来临。若是挑战极限，还可营造北极冰寒的效果。

REFRESHING   清凉酷爽

We recognise that this is the colour world which is the spitting image of the world of the soft drinks and mineral water industry. Sparkling air bubbles rise and establish the closeness to that sector.

这些颜色所呈现的是软饮料之海、矿泉水之海，尤其是那些亮晶晶的炫彩气泡。

VITALISING

无限活力

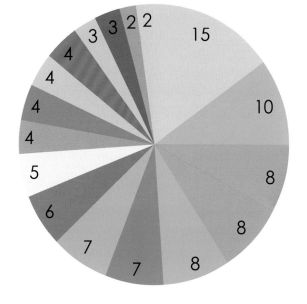

Predominantly dynamic and cheerful! The distinctive turquoise and ocean-blue shades are widespread and occupy about forty percent of the chart. The lime-yellow-green shades follow with 25 percent.

映入眼帘的是活力十足、赏心悦目的色彩组合。与众不同的松石绿和海蓝占据很大比例，约达40%。其次是浅绿和黄色，占25%。

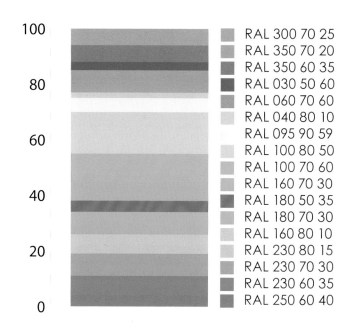

The stripes bear a strong resemblance to the colour schemes of Hawaiian shirts. This colour sequence livens up even the exhausted.

色条让人不由自主地想到夏威夷衬衫的缤纷多彩。充满活力的色彩足以让筋疲力尽的你瞬时抖擞起精神。

The number of interpretations sketched is very low. For the study participants, the startling and arousing features of the colours were distinctive enough.

这一风格的典型图案并不多，在研究参与者看来，这些令人目眩的色彩组合就足以使它独具魅力。

VITALISING 无限活力

Animated with summer hues, jungle greenery and wild water as well as vibrant garden colours, these shades give strength and energy to the beholder. The drawing style is mainly flat and unstructured.

这里有夏天的炽烈色彩，如热带丛林、湍流急水和生机盎然的花园，放眼看去就能感受到满满的能量。线条平实而不拘一格。

APPETISING

A look at the fruit stand on the market in the middle of the summer. A tasty fruit pie chart has been prepared, which magically attracts those with a sweet tooth.

秀色可餐

仲夏时节集市上的各色水果，便是"秀色可餐"风格的色彩映照。果色生香的诱惑，让爱吃甜食的人们无法抗拒。

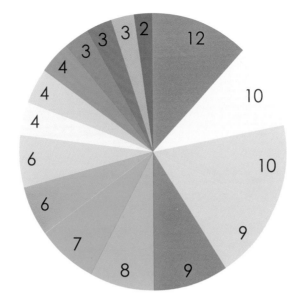

The colour stripes convince us that "appetising" is an adjective with an especially effective quality: the expectation of a foretaste is irresistible, and anticipation is as great as the actual enjoyment is healthy.

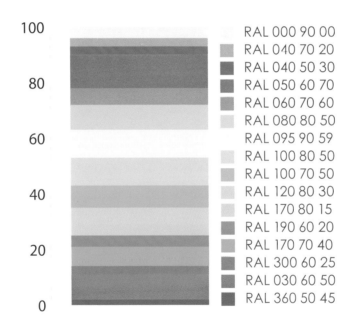

RAL 000 90 00
RAL 040 70 20
RAL 040 50 30
RAL 050 60 70
RAL 060 70 60
RAL 080 80 50
RAL 095 90 59
RAL 100 80 50
RAL 100 70 50
RAL 120 80 30
RAL 170 80 15
RAL 190 60 20
RAL 170 70 40
RAL 300 60 25
RAL 030 60 50
RAL 360 50 45

丰富多彩的色彩条再一次证明了"秀色可餐"的强大诱惑力：美食面前谁也不能自持，对美食的期待与享用美食的过程，都让人无比享受。

Signs come off badly. The colour impression suffices as a visual message.

图案的风格看起来差强人意，但强烈的色彩效果足以让人们忽略图案的弱势。

APPETISING　　　　秀色可餐

Like freshly prepared crudités with summer fruits of all sorts, very stimulating, tasty, ranging between sour and sweet. That is what appetising colours look like. No black, no light blue, no vibrant blue has gone astray in this colour range.

刺激味觉的色彩大餐，如夏季水果组成的新鲜沙拉般酸甜可口。拒绝黑色、浅蓝和鲜蓝，还原"秀色可餐"的诱人本色。

HEALING
情感治愈

REFRESHING
清凉酷爽

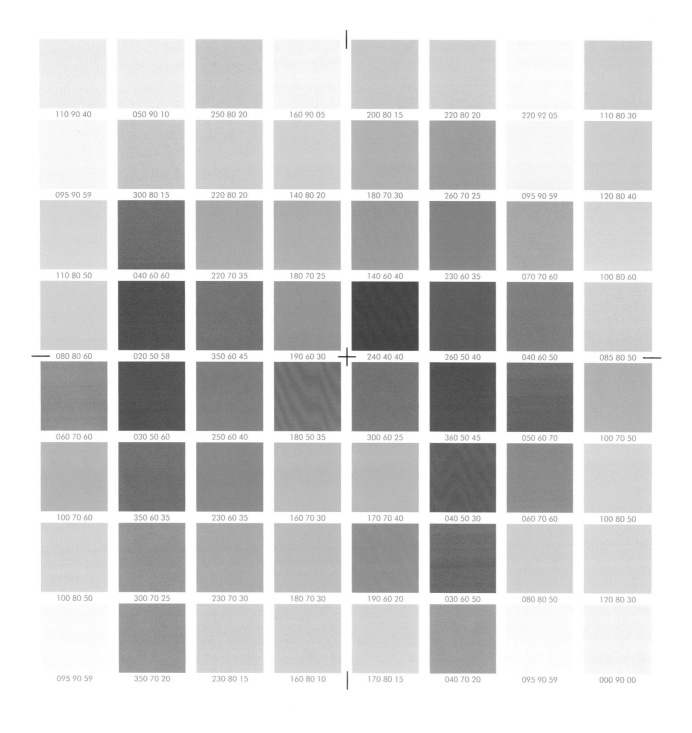

VITALISING
无限活力

APPETISING
秀色可餐

HEALING
REFRESHING
VITALISING
APPETISING

情感治愈
清凉酷爽
无限活力
秀色可餐

The four colour ranges feature activating characteristics. Their appeal aims for stimulation and excitement; they are reminiscent of heraldry, as if in a battle of the banners of the most beautiful colours. It is generally the pure colours, not those covered in dust and dirt, that trigger animating processes. What here looks quite colourful and uncoordinated gains a reality that is sensible, purely theme related and thus convincing, focused on emotion and understandable, through a restrictive colour selection.

Warm, bright cream and sorbet hues should be used predominantly as colours for large surfaces. However, the active colours, especially red nuances and, occasionally, a few distinctive blue-green shades, are very effective when it is a matter of a sensible and pleasant design approach that promotes health and wellbeing. Of course the light hues on the outer edge of the colour chart have a positive creative power, too, if they are applied on large areas in the context of medium shades.

这四种风格的颜色都很鲜明。它们刺激感官、点燃激情；它们遵循纹章学，像战场上飘扬的旗帜一样色彩鲜明。选色主要为纯色，洗去尘埃与铅华。多样化的色彩组合看似不够协调，却是对现实世界的理性映照。通过对颜色的精心搭配，更能触及顾客的真实情感，引起心灵共鸣。

温暖鲜亮的奶油冰糕色调占据了大部分的比例。不过，活跃的色彩比如红、蓝和绿色，对于营造明快愉悦、充满正能量的空间氛围，同样效果惊人。还有表格外围的浅色系，如果运用得当，比如为它们搭配中性色的背景，也将碰撞出无穷的创造力。

VITALISING

无限活力

100 70 60   100 80 50

We perceive as constructive and healthy shades that look as if they have been copied from fruit juice drinks. Only bright, fruit-coloured shades suggest impressions of freshness, liveliness, performance and receptivity. This fact can hardly be related to the well-known placebo effect, because colours convey positive (and occasionally also negative) qualities to us, the sensory and emotional capacities of which we also process as similar kinds of neuronal mood boosters when we experience a beautiful, sunny day or see a fantastic bunch of flowers.

100 70 60   100 80 50   350 70 20

Green carpets in guest rooms and green walls promote freshness, especially if the complementary rose-red emphases generate additional emotional enrichment.

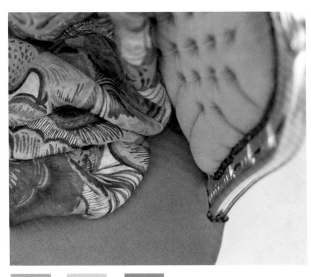

160 70 30   100 80 50   350 60 35

我们不难发现，积极、健康的色调往往以果汁饮料作为参考源。鲜亮的水果色便象征着新鲜、活力、能量和超高的受欢迎度。这一事实与心理咨询领域的"安慰剂效应"完全无关，因为色彩传达给我们积极的（偶有消极的）感知特征，如同我们在阳光明媚的一天看到满园春色一样，因心情舒畅而获得健康身心。

The design processes feature many details: this includes well-placed fruit bowls and fresh greens, sometimes presented in an over-abundant and voluminous manner. Again, large wall surfaces, beautifully draped fabrics with floral colours and drawings are important. Due to its conspicuous and unfamiliar colourfulness, an agate-green upholstered antique settee creates a great attraction.

Most of those who sit on green cushions for the first time are surprised to find how wonderful and relaxing this can be. And of course this is also a useful hint for more relaxed driving!

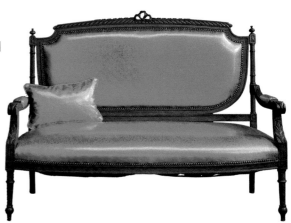

160 70 30   230 70 30   350 60 35

180 50 35   350 60 35   100 70 60   250 60 40

酒店客房的绿色地毯和绿色墙壁，营造出充满新鲜感的氛围，再以玫瑰红作为搭配，让空间的情感体验变得更加丰富。

设计过程中需要充分留意细节：比如水果盘和绿植的正确摆放，有时需要把它们组成单元集中布置；还有对大型墙面的装饰、对织物表面的花色或绘图的选择等，虽是细枝末节却不容忽视。由于空间的用色新奇大胆而且格外丰富，一张单色系的玛瑙绿古董沙发反而成为视觉焦点。

第一次将身体陷入这绿色软垫，才发现那是怎样美妙的享受！无法想象，如果把它作为汽车坐垫，又会带来何等舒畅的驾驶体验！

250 60 40

350 70 20

350 60 35

100 70 60

040 80 10

300 70 25

060 70 60

100 70 60

100 80 50

095 90 59

HOTELS – WITH A HOST FAMILY: VILLAGE-LIKE TO PLUSH

If we have not yet encountered these hotels, we should make a point of filling in this gap in our experience. If the opportunity arises, be it in the countryside, in a village or a small town, away from airports, motorways and the screeching of a goods train, we should seize it. All the more so if the hotel is situated on a village road and bears the patina of a traditional inn.

In hotels like these, few guests stay in their room for any length of time after arriving. Going downstairs to the lounge or the bar serves not only to quench one's immediate thirst, but perhaps also to revive old acquaintances or, if it is the first visit, to make the acquaintance of the landlady or the landlord. We soon discover that in many cases the remnants of the splendour of an old coaching inn have been preserved. The first questions and answers will concern the whence and the whither, one's well-being and further travel plans, with enquiries as to thirst and hunger. There follow particulars of the specialities that the kitchen and cellar have to offer.

Ancestors and sometimes other relatives hang on the walls in sedate picture frames; group portraits with ladies, children, fathers, grandfathers and great-grandmothers, often in local costume, can be admired if the pictures are more than one hundred years old. After half an hour, guests, landlady and landlord have been informed about the family backgrounds of both families. Occasionally you may even meet the exuberant grandchild, the future patron.

These delightful hotels and guesthouses have many facets, which are described by means of several examples. When analysing our studies we discovered that the design and characteristic features are clearly differentiated. For example, the village hotel image in frequently earthy, woody and herbal tones can be distinguished from the purely family-friendly hotel type by its use of warm, vernal and summery colours close to nature. The colour scheme of the tourist model is considerably cooler, whereas the ambience of the homely hotel requires well-behaved, muted shades of brown, warm chromatic colours and hints of sky blue. The country hotel is described with more vegetal colours than the robust, rustic type.

The guests love the charm of small, medium-size and even large hotels in the countryside because the rhythm of the time one spends in the hotel inn moves at tractor speed and not in a flash. It is pleasant to stay in a folkloristic ambience that remains closely in touch with its native soil, to interact with a vernacular colouring and taste local gastronomic pleasures. As cosmopolitans we are learning increasingly to appreciate the rural-folkloristic scenario, even if it is different from the familiar homeland attributes.

家庭式酒店：
从乡村风格到豪华舒适

未曾遇到此类酒店的人们应填补这一空缺。如果有幸在远离机场车道、火车鸣笛的村庄或小镇遇到这种酒店，我们应当珍惜它。倘若我们更加幸运，该酒店恰好位于乡间小道旁，还有着传统酒馆的气息，那就更不能错过了。

入住此类酒店的人们很少会一直待在房间里，而是喜欢到楼下的休息室或酒吧中，或饮酒解渴，或重叙旧情。若是初次来到此处，还会去结识房东。往往很快我们就能感受到旧酒馆的余辉。人们首先会思考的问题就是"从哪儿来，到哪儿去？"答案也都关乎此行的感受和旅行的下一站，还会收到各种询问是否饥渴的寒暄之词。许多细节之处都关乎厨房和酒窖，而此类酒店恰好发挥了这两者的优势。

祖先或者其他亲戚的肖像被稳重的相框框起来，挂在了墙壁上；有的群体画有逾百年的历史，画中都是身着当地服饰的女人、孩子、父亲、祖父和祖母，值得欣赏一番。半个小时以后，客人和房主已互相了解家庭背景，甚至有时还见过了多名曾孙和继承人。

这类酒店和旅馆都令人感到身心愉悦，却也各有特色，以下将用几例来说明。在研究中我们发现，它们在设计和特色方面大相径庭。如乡村酒店往往以大地色、草木色为主，而家庭旅馆则以柔和、有春夏气息的色彩为主；以旅游为导向的酒店用色清淡，而家庭式旅馆的良好氛围往往是通过暗棕色、暖色和些许天蓝色来营造的；相对粗犷淳朴的酒店而言，乡村酒店往往采用的植物色较多。

顾客往往难以抗拒中小型乡村酒店的魅力，甚至是乡村大酒店也颇受大众喜爱，因为酒店的生活节奏很慢，无须忙碌。感受酒店的乡村气息，了解乡土民俗，学习当地语言，品尝当地美食，会令人心情舒畅。作为都市人的我们，越来越懂得如何去体味乡间故事，再现民间故事中的场景，即使这些与自己熟悉的家园截然不同。

164	village-like - natural - authentic - down-to-earth	乡村风格 – 回归自然 – 返璞归真 – 朴实无华
180	restrained - strict - simple - sober	婉约矜持 – 一丝不苟 – 简约朴素 – 素净质朴
194	family-friendly - touristic - homely - appealing	居家温情 – 爱在旅途 – 家常记忆 – 新奇有趣
206	rural - secluded - rustic - puristic	山水田园 – 与世隔绝 – 原始乡土 – 纯粹至美
220	historic - antique - feudal - plush	历史沉淀 – 古色古香 – 盛世王侯 – 豪华舒适

/ Chapter 3

VILLAGE-LIKE

乡村风格

All the colours of field, forest, garden and meadow, sky, weather, autumn, summer, and spring are assembled here. A panorama of village life and colour structures.

这里集合了各种乡村色彩，如田野、森林、花园、草地、天空、风雨云雪、春夏秋冬。这就是乡村生活的全貌、乡村色彩的所有。

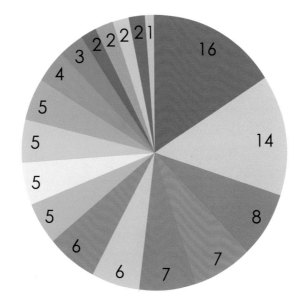

Emulation recommended! Wish and cliché, sentiment and reality present a concept that comes very close to the authentic.

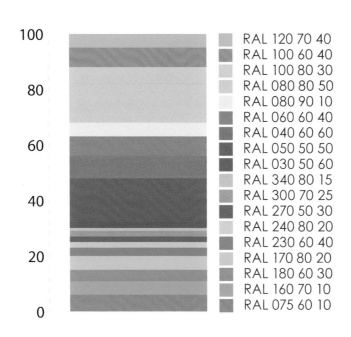

RAL 120 70 40
RAL 100 60 40
RAL 100 80 30
RAL 080 80 50
RAL 080 90 10
RAL 060 60 40
RAL 040 60 60
RAL 050 50 50
RAL 030 50 60
RAL 340 80 15
RAL 300 70 25
RAL 270 50 30
RAL 240 80 20
RAL 230 60 40
RAL 170 80 20
RAL 180 60 30
RAL 160 70 10
RAL 075 60 10

这些乡村色彩值得推荐，期待与陈腐、感觉与现实，这种对比让理念更加接近真实。

Symbols of the landscape, of architecture, flora and earthy, stony materiality are recognisable in the drawing style.

用绘画笔法绘出来的这些图案，不难看出表现的是乡村风景、建筑、植物和土石。

VILLAGE-LIKE   乡村风格

This is the exact representation of a Central European village idyll. Its abstraction is incomparably harmonious and corresponds to the feelings and expectations of those looking for recreation.

这恰似一幅中欧乡村田园画，抽象而又和谐，反映了寻求放松时刻的人们的感受和期望。

/ Chapter 3

NATURAL

回归自然

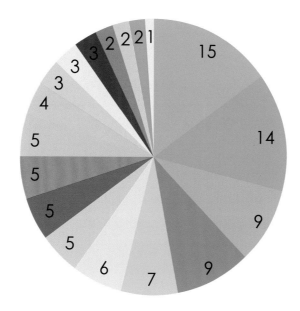

Things are growing and thriving. Vegetable colour models, but also wind, weather, sun, woods, and earth characterise the colour picture.

色盘中呈现出万物复苏、欣欣向荣的景象，蔬菜的色彩、风雨云雪的色彩、阳光和树木的色彩、大地的色彩都囊括其中。

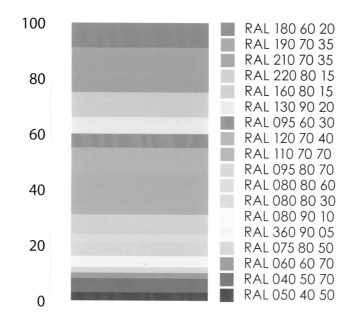

RAL 180 60 20
RAL 190 70 35
RAL 210 70 35
RAL 220 80 15
RAL 160 80 15
RAL 130 90 20
RAL 095 60 30
RAL 120 70 40
RAL 110 70 70
RAL 095 80 70
RAL 080 80 60
RAL 080 80 30
RAL 080 90 10
RAL 360 90 05
RAL 075 80 50
RAL 060 60 70
RAL 040 50 70
RAL 050 40 50

The natural range is the photographic portrayal of a vernally influenced Central European landscape with forests and meadows.

这些自然中的色彩组成了一幅中欧风景画，春风拂过，花草树木都在炫耀自己鲜绿的色彩。

Signs of still waters, burgeoning plants, rainbows, and colourful flowers: idyllic and close to nature.

静静流淌的河水、生机勃勃的树木、五光十色的彩虹、缤纷多彩的花朵，这些都是田园的元素、大自然的造物。

NATURAL 　　　　　　　　回归自然

The most natural thing in the world is predominantly green. Starting with unripe lemon green to Mediterranean water green. As complementary shades, a few sky-blue nuances and several withered looking blossom hues can be discovered.

从嫩嫩的柠檬绿到地中海海水般的碧绿，世界上最自然的色彩莫过于绿色。再微微掺入些天蓝色调、浅浅的花朵色，便完美无瑕了。

/ Chapter 3

AUTHENTIC

返璞归真

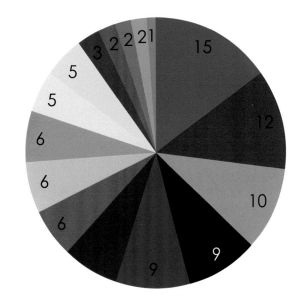

The "authentic" can be recognised by its unaffected metal and earthy shades in bold and deep-toned hues.

这里说到的"返璞归真",可以通过色彩鲜明、色调更深却又无华的金属色和大地色来表现。

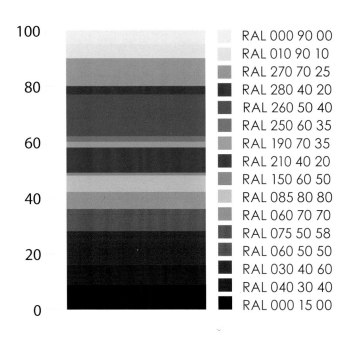

RAL 000 90 00
RAL 010 90 10
RAL 270 70 25
RAL 280 40 20
RAL 260 50 40
RAL 250 60 35
RAL 190 70 35
RAL 210 40 20
RAL 150 60 50
RAL 085 80 80
RAL 060 70 70
RAL 075 50 58
RAL 060 50 50
RAL 030 40 60
RAL 040 30 40
RAL 000 15 00

The bright hues at the top edge of the graph show the trivialisation of the authentic. The hues at the bottom, however, confirm it.

色卡顶端的明亮色彩体现了朴实的内涵,底端的暗色调更是证明了这一点。

The drawn examples of the authentic are graphical, rational and simple.

与这类色彩相对应的是简单、规则、明确的图形。

AUTHENTIC 返璞归真

The authentic is first and foremost static, unambiguous and sure. The original gets by without major grass-roots grumbling. The content of realisation and recognition.

"返璞归真"的最大特点就是沉静、明晰、真切。挣开本性的禁锢，追逐新生，这就是成功与被认可的实质。

/ Chapter 3

DOWN-TO-EARTH

朴实无华

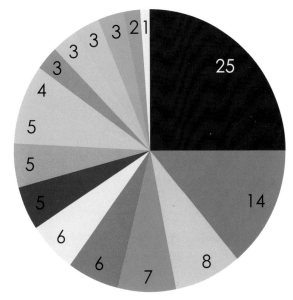

The factual is difficult to endure and not always bearable if it is interpreted as technoid. In our imagination, everything that is factual refers to reality in shades of black and grey.

若现实主义被定义为technoid，则这种手法将很难总是得到认可并长久存在。我们所想的现实主义，应该更偏向于一种灰黑色调。

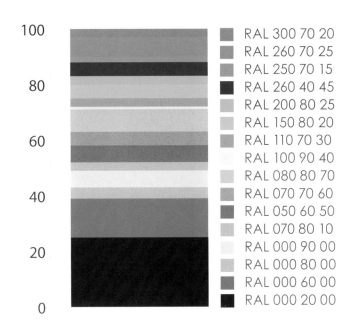

- RAL 300 70 20
- RAL 260 70 25
- RAL 250 70 15
- RAL 260 40 45
- RAL 200 80 25
- RAL 150 80 20
- RAL 110 70 30
- RAL 100 90 40
- RAL 080 80 70
- RAL 070 70 60
- RAL 050 60 50
- RAL 070 80 10
- RAL 000 90 00
- RAL 000 80 00
- RAL 000 60 00
- RAL 000 20 00

The colour range is reminiscent of primeval times, which were martial; we do not know if they were particularly gloomy. At least, we refer to it only as grey but not as black.

这组色彩会令人联想到遍地争战的原始时代。那时是否特别阴郁黑暗，我们并不知道。至少，我们用灰色而不是黑色来指代阴郁。

Symbols from years of devotion to technology. Constructive symbols characterise the style.

这些构造清晰的符号来自探索科技的时代，明确地说明了这种风格的特征。

DOWN-TO-EARTH 朴实无华

Here we see the impression of the industrialisation of kitchen or living areas, designed by a function-driven engineering caste. Possibly only those who have been infected with steel and iron microbes will feel comfortable here.

仔细品味上图，我们会联想到工业化的厨房或生活区，其内部由一个功能驱动型工程覆盖，或许只有被钢铁病菌感染的人们才能在这种环境中感到舒适。

/ Chapter 3

VILLAGE-LIKE
乡村风格

NATURAL
回归自然

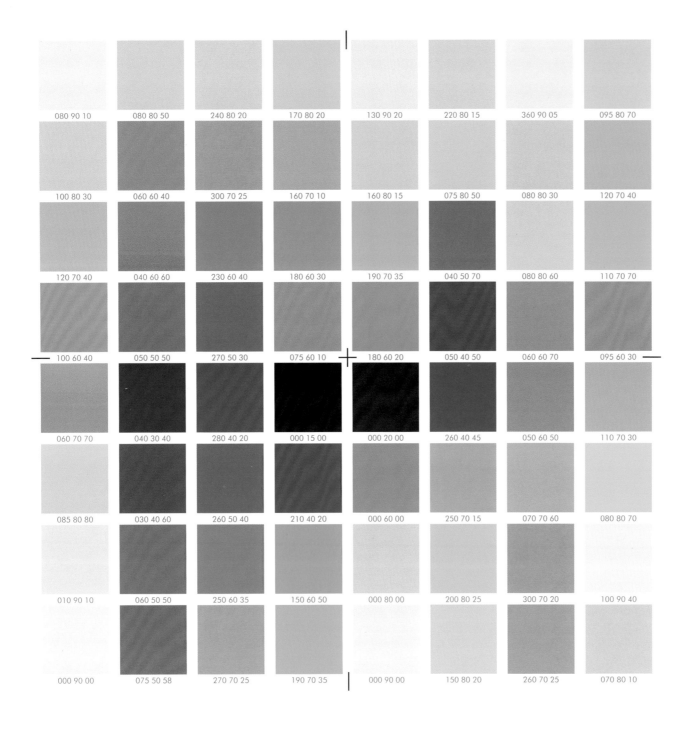

AUTHENTIC
返璞归真

DOWN-TO-EARTH
朴实无华

VILLAGE-LIKE NATURAL AUTHENTIC DOWN-TO-EARTH

乡村风格
回归自然
返璞归真
朴实无华

The emphasis of the terms lies on honest foundations that have evolved over time. These four theme and colour worlds demonstrate an unadorned unpretentiousness and a contact with reality: real life beyond hectic, noise and too much artificiality. This is what we call upon if we take up residence away from main roads and the hustle and bustle. Of course, we will re-discover many things here that we declared we could live without only a few years ago. At intervals, our own history keeps catching up with us. How soothing are the bulgy tables and side tables, the cute bag lamps and the high-pile carpets for slouching in front of the fireplace!

The results of our colour research looked similar ten or even twenty years ago because our collective experiences are almost identical: plenty of red-brown, bright and dark green, mid-and dark blue, and a few white-and limestone-coloured variations. Materials and surfaces we have been familiar with for years.

这些词语强调的是，随着时间的推移而形成的稳定的本质。这四种主题和色系都展现出了质朴之美以及与现实的联系：真实的生活应远离尘嚣，摒弃矫揉造作。这便是当我们想要逃离车水马龙、熙熙攘攘的城市时想要的生活。过上这种生活，我们能够对诸多几年前才被生活淘汰的物什进行探索和发现。此外，关于我们自己的故事也不断在脑海中浮现。想象一下，凹凸不平的桌子和茶几、可爱的袋状灯、舒适的毛毯以及暖和的壁炉，身在其中是如此舒适。

此次色彩研究的结果与十年前甚至二十年前的研究结果相似，因为我们有着几乎相同的经历。这些色彩包括丰富的棕红色调、亮绿和深绿色调、深蓝和中蓝色调、些许白色和灰色调。拥有这类色彩的材料和外观我们都早已耳熟能详。

AUTHENTIC 　　　　　　　　　　　返璞归真

We perceive what is original and credible to be authentic. We use the term to describe the feeling and the certainty that everything must be genuine and appropriate for the material used: real leather, real fur, real wood, real stone, real parquet, real table linen with real napkins, and real cosiness. We noticed that reality has nothing to do with its media appearances and that it is determined by illusion rather than reality.

It is adventurous to sit at a table which is not made from chipboard but from real wood, when the width of the logs which were used can still be recognised. Our relationship to products which originate from living nature are a great deal more stable and emotive than they are to those made from fake stones, earths and ores. Mostly, they are too cold, too hard, too rugged. If the armchairs in the fireplace room were covered with white or black leather, they would appear to us less quaint than the ones in shades of cognac-brown.

060 50 50   010 90 10

　　我们通过感知来辨认什么是源、什么是真，并将其总结为"返璞归真"（authentic）一词。我们用这个词来描述我们笃定的感受，认为关乎材料的一切都必须是真的、恰当的，如真皮、真毛、实木、真石、实木地板、真餐布和真餐巾，还有一种真实的舒适感。然而我们发现，事实与外观媒介似乎没有太大关系，这种真实感更多的是来自一种假象，而非现实。

210 40 20

| 060 50 50 | 040 30 40 | 000 15 00 | 000 90 00 | | 250 60 35 | 080 90 10 | 270 70 25 |

| 075 50 58 | 010 90 10 | | 010 90 10 | 000 90 00 | 060 50 50 |

The most attractive meeting points in the hotel are always the ones where people can see each other. These are the archaic-authentic rooms of social exchange, of getting to know each other and of making friends.

The hotel as an interpersonal meeting place needs such areas. In a neutral space so to speak, they make great socio-economic sense. Frequently, they facilitate the initiation of profitable relationships which are established outside the office. And no ambience is better suited for this purpose than one which was designed close to nature with a combination of ambitious originality and authenticity. Here, everything is done correctly: the armchairs are massive and large and ensure a sufficient distance to the neighbour and the person sitting opposite.

010 90 10   075 50 58   000 90 00

010 90 10   075 50 58   060 50 50   000 90 00

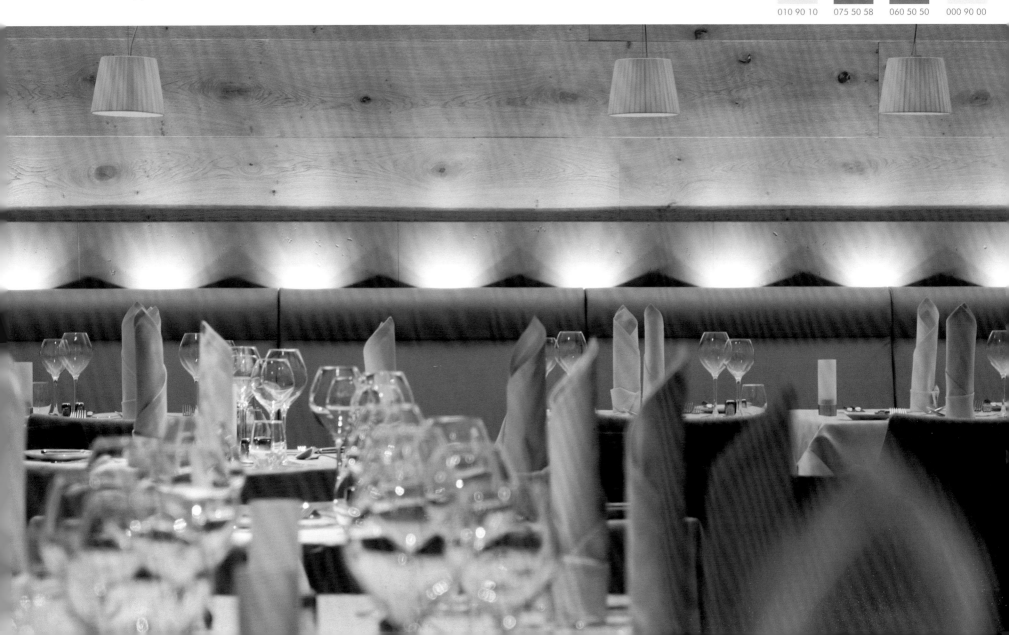

040 30 40

075 50 58

010 90 10

060 70 70

坐在实木桌而非由木屑板做成的桌子旁时，人们会惊奇不已，特别是发现桌上的木纹还清晰可见时，更会感到惊讶。相对假石、假土而言，我们往往更钟情于源自自然的物品，毕竟这些假物都太冰冷、生硬了。试想一下，壁炉前摆放的扶手椅若采用的是白色或黑色的皮革，则显然没有使用棕色皮革来得亲切。

000 15 00   040 30 40   060 50 50

酒店中最有吸引力的约会地点往往是相对开放、视线无阻的地方。这些古朴的空间适合人们会谈交友、相互了解。

酒店作为交际会谈的场所应当设有这类空间，例如中性的空间往往有很大的社会经济价值。通常，互利共赢的合作关系都是在办公室外建立

000 90 00　　000 15 00　　075 50 58

的，而在自然、新奇、真实的空间氛围中，这种关系最有可能达成。此空间中的一切布置都十分合理，沙发舒适宽敞，座位之间距离恰当。

060 50 50

075 50 58

000 15 00

000 90 00

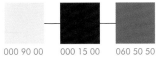

000 90 00　　000 15 00　　060 50 50

/ Chapter 3

RESTRAINED

婉约矜持

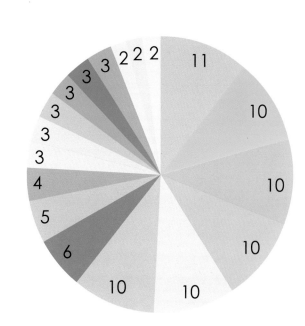

The range is made up of highly differentiated limestone and cement colours lightened by pale lemon shades as well as sky blue and cloud hues.

色盘中石灰色和水泥色居多，并有淡柠檬色、天蓝色和云彩色来增添生气。

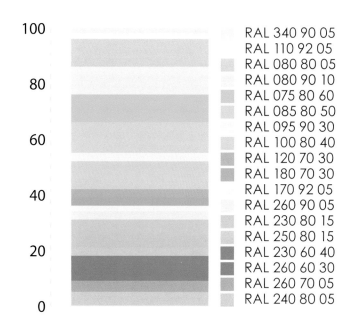

RAL 340 90 05	
RAL 110 92 05	
RAL 080 80 05	
RAL 080 90 10	
RAL 075 80 60	
RAL 085 80 50	
RAL 095 90 30	
RAL 100 80 40	
RAL 120 70 30	
RAL 180 70 30	
RAL 170 92 05	
RAL 260 90 05	
RAL 230 80 15	
RAL 250 80 15	
RAL 230 60 40	
RAL 260 60 30	
RAL 260 70 05	
RAL 240 80 05	

The sequence looks exactly as it is labelled. We can recognise a reduced elegance and visible denudation of biological substance.

这组色彩正如其名，内敛婉约，就像是被侵蚀的生物体一样，其精美之处仍依稀可见。

The interpretation of drawn details remains very scanty: sometimes a circle, a few lines, otherwise nothing but foggy patches and neutral areas.

四幅图案虽笔画稀疏，但细节之处值得关注，如似圆非圆的图形、断裂的线条、整片模糊或部分模糊的区域，都充分阐释了婉约的含义。

RESTRAINED　　　　　　　　　婉约矜持

Meagre colours were largely selected. Accordingly, a significant number of Cinderella shades have turned up. They look pallid and pale rather than energetic. If well combined, they offer shades of aristocratic pallor.

图中非常浅的色彩居多，还有大量的灰色，就像是灰姑娘的装束一般，十分朴素。这些色彩看起来暗淡、无活力，但如果搭配恰当，会给人一种贵族般的威严感。

/ Chapter 3

STRICT

一丝不苟

The pie chart makes it clear: the dark shades, which are accompanied by a few contrasting hues, emphasise the deep-toned foundation.

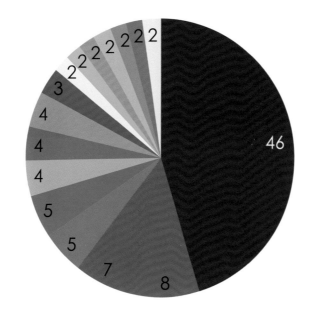

此扇形图直观地体现了"严肃"色系的代表色就是黑色调，其他几种对比色调的掺入，更加突出了深色的主色调。

The black surfaces and the coloured lines mark the claims of the unapproachable but stable, protective and monumental block of power.

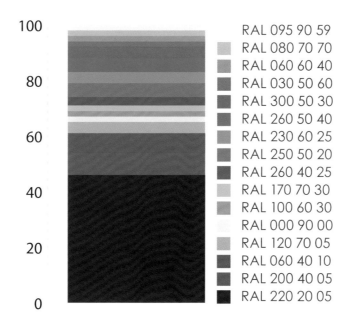

RAL 095 90 59
RAL 080 70 70
RAL 060 60 40
RAL 030 50 60
RAL 300 50 30
RAL 260 50 40
RAL 230 60 25
RAL 250 50 20
RAL 260 40 25
RAL 170 70 30
RAL 100 60 30
RAL 000 90 00
RAL 120 70 05
RAL 060 40 10
RAL 200 40 05
RAL 220 20 05

大片黑色区域和其他彩色条组合在一起，体现出一种难以亲近却沉稳、自我保护且不可撼动的力量。

The appropriate features of this motionless expression are graphic, linear and monolithic.

单调的直线或图形恰到好处地诠释了这种稳如泰山般的感觉。

STRICT

一丝不苟

Architectural stringency is expressed by the colour field made up of forty-nine individual images. It does not come as a surprise that black and grey values dominate the overall impression. The same applies to the scattered yet precise colouristic and graphic corroborations.

由49张图案组成的色彩图说明了建筑的严密性。黑色和灰色主导了整幅色彩图，当然这不足为奇，更引人注目的是分散而又精致的彩图。

/ Chapter 3

SIMPLE

简约朴素

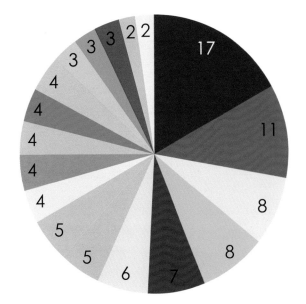

The colour range reflecting apparent scarcity and a chilly climate with the basic shades is slightly vapid and overly plain. Here, it is mainly a question of targeted combinations of colour and shape.

由过于单调朴素的色彩组成的色盘给人一种走进寒冬的感觉。其中的主要难题就是色彩和形状的组合问题。

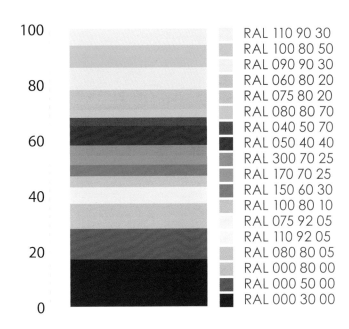

Unfortunately, no beautiful magic appears with this configuration. The dread of simplicity and forced renunciation remains succinctly present.

这组色彩中并未藏有美丽的魔法，只有令人畏惧的单调感和荒芜感，或许你会因此感到惋惜。

Various indications of a grid, lines and chaotic brushstrokes.

网格、线条、混乱的笔画似乎有多种含义。

SIMPLE   简约朴素

The first glance proves: that's exactly what some simple guesthouses look like and we occasionally quite like to be reminded of them. A bit gloomy, a bit too yellow, then a faded red, a loud green, and a plastic-coloured light violet.

一丝暗色、些许昏黄、褪去的红色、醒目的绿色以及淡淡的紫色，让人看一眼就会觉得这正是简约客房的样子，有时回忆这种酒店内的场景都会令我们感觉很享受。

/ Chapter 3

SOBER

素净质朴

Encounters with such shades are not encouraging. They are no more than factual hints to replace bright colours with the plainest of hues.

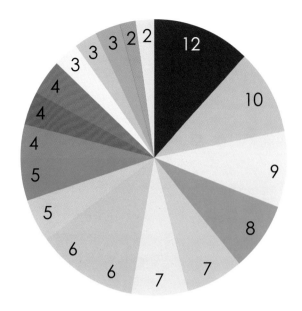

与这些颜色"邂逅"着实不能令人兴奋。它们只不过是用最朴素的色调替换鲜亮色彩时留下的真实痕迹。

Mind you, the colour sequence appears neat and tidy. It makes the beholder an austere witness of a reduced reality.

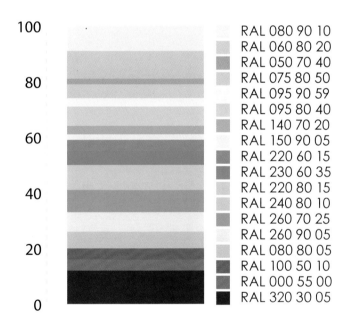

RAL 080 90 10
RAL 060 80 20
RAL 050 70 40
RAL 075 80 50
RAL 095 90 59
RAL 095 80 40
RAL 140 70 20
RAL 150 90 05
RAL 220 60 15
RAL 230 60 35
RAL 220 80 15
RAL 240 80 10
RAL 260 70 25
RAL 260 90 05
RAL 080 80 05
RAL 100 50 10
RAL 000 55 00
RAL 320 30 05

请留意,这一颜色序列表现得非常简洁有序。它能让色彩的载体成为残酷现实的见证。

A few lines and otherwise only swirling or nebulous amorphous shapes representing something blurred.

要么是几根简单的线条,要么是不规则的漩涡,抑或是模糊的星云状图案,用以代表一些模糊的事物。

SOBER 素净质朴

Colours one does not look consciously at and a handful of colour accents one hardly notices. These meagre impressions indicate the abandonment of cheerful or emotional colour pictures.

倘若不是有意去观察，人们不会注意到这些色彩，因为它们都不醒目。这些素雅的色彩未带任何感情色彩，不会令人感到振奋。

/ Chapter 3

RESTRAINED
婉约矜持

STRICT
一丝不苟

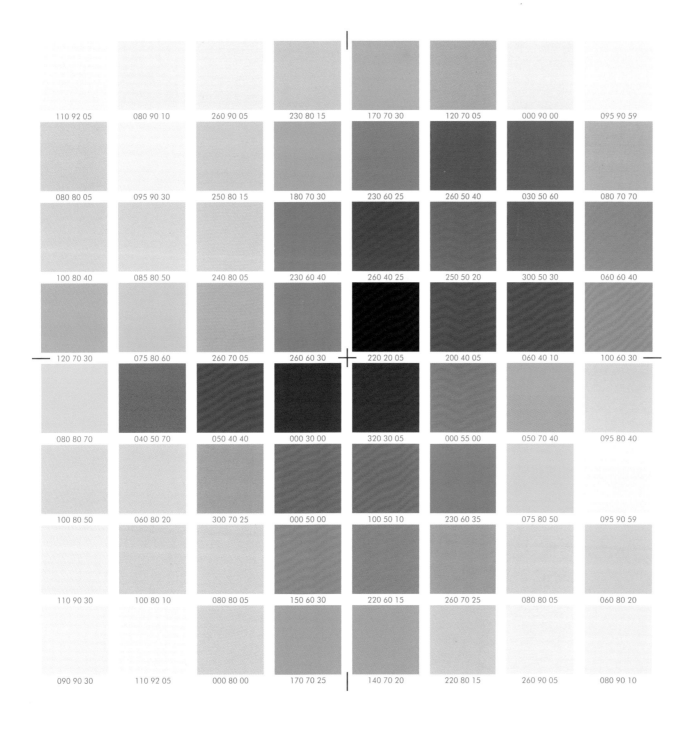

SIMPLE
简约朴素

SOBER
素净质朴

RESTRAINED
STRICT
SIMPLE
SOBER

婉约矜持
一丝不苟
简约朴素
素净质朴

These pages illustrate that something sensible and appealing can be designed with a clever, imaginative, creative interpretation.

Sometimes, the orientation towards the unspectacular, strict and factual is an absolute necessity. This is how we feel after three or four weeks of heat and blazing sun, when we experience rain and a fresh breeze as a relief. Even when we notice that the table we are sitting at is made of solid beech wood, we learn to once again appreciate the original.

We will surely appreciate that the simple chair made of beech wood has a solid base and impresses us precisely because of its design austerity. As in everyday life, we are increasingly fond of the original after an excess of over-elaborate decoration, just as we prefer plain, local dishes as opposed to those that smell of over-refined flavours.

以上几页内容表明，用充满智慧、想象和创造力的手法可以设计出清晰明朗、充满魅力的作品。

有时，很有必要将风格定位为平实、精确、真实。这就是为何在高温炎热天气过后，我们需要大雨和凉风的慰藉。即使我们发现身旁的桌子是由山毛榉木做成的，也仍有必要去仔细观察其天然纹理。

简单的山毛榉木椅会令人喜爱不已，其上的实木基座设计如此简单，更是令人大为赞叹。在日常生活中，我们习惯了过分精致的装饰物品，因而越来越偏爱原始自然的物品，比如，我们会更喜欢原汁原味的地域菜肴，而非口感细腻的佳肴。

RESTRAINED 婉约矜持

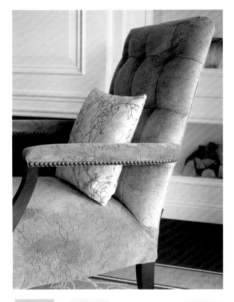

080 80 05   340 90 05   110 92 05   080 90 10

Restraint also is the most elegant form of understatement. Frequently, its value only makes itself felt at second glance. We often admire people for these qualities. Mostly it is those people who have a lot to say and are only prepared to make their opinion heard when they consider it prudent to do so. They are the truth-loving counterpart to impulsive chatterboxes. There is nothing convoluted and loud about them. This furnishing scenario presents itself in precisely this manner: there is no splendid, theatrical scenery, but it represents a contemplative, well-tempered, interior design determined with a view to relaxation.

Colours, shapes, surfaces and silhouettes generate quiet, attractive images promising the user calm and recreation. Anything soft, arches and circular shapes are frequently used ingredients of this pleasantly casual atmosphere because nothing is artificial or inconsistent; everything is based on a synthesis of similar and coherent design features. People also like the language of the materials, their natural colouring and thus their concomitant honesty. Trying to create a false impression is frowned upon, as are dried flowers, large mirrors, and anything faux and fake.

The functional in the sense of design honesty is at the top of the list. Design needs to be interpreted as function. Design also needs space and no restricting circumstances. And it needs order, direction and hints for finding the way. Everybody has already taken a wrong turning through many a hotel. Sometimes they are chaotic and confusing, and the direction indicators are hidden and written in print that is far too. That is not only senseless but annoying, because it faily to regard or even neglects the cognitive ability of the clientèle. In any case, space-consuming clarity is better than space-saving narrowness.

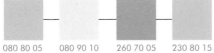
080 80 05   080 90 10   260 70 05   230 80 15

080 80 05   080 90 10   110 92 05

080 80 05   340 90 05   080 90 10

260 70 05   260 90 05   110 92 05

The cubic interior design of the wall, floor and window areas, of the upholstery and the entire furnishing through to the technology of switch elements make up the main part and the prime attributes of the scenery. Even the zebra elements of the carpet join in.

110 92 05   340 90 05   080 90 10   260 70 05

085 80 50   110 92 05   095 90 30

朴素而又优雅，则为婉约。这只有反复品味，才能有所体会。朴实低调的人总是令人钦佩，他们怀有许多见解，然而只有在深思熟虑之后，才会挑选恰当的时机表达出来。与鲁莽的"话匣子"相比，他们更热爱真理，在遇到难题时也总能保持头脑清晰，因此难题也就迎刃而解了。图中的装饰恰好也有这种风格：没有富丽堂皇的装饰，没有夸张的氛围，只有沉静与平和，令人身心放松。

色彩、形状、表面和轮廓相互作用，组成安静迷人的画面，令身处其中的人感到平静和愉悦。这种氛围的塑造得益于线条柔和的圆形物品的使用，空间内一切物品有着某种共同点，从而显得自然、和谐。人们也会喜爱这种用材方案、自然的色彩以及伴之而生的真切感。绞尽脑汁制造的假象往往会令人心生厌恶，如干花束和大镜子等人造假物。

"忠实"型设计的重点是实用性。设计要与功能相结合，要有发挥的空间和自由，更要有规则、方向和暗示来帮助寻找路径。或许每个人都曾在酒店中被误导过。因为许多酒店的设计并不合理，并没有考虑甚至有意忽视了客户的认知力，从而显得混乱不堪，导视标志也并不直观，甚至是打印的，令人对此毫无好感，甚至觉得不快，在任何情形之下，通透大方的空间都比狭小紧凑的空间更优越。

110 92 05

340 90 05

240 80 05

260 70 05

260 90 05

方形的墙壁、地面和窗户，以及方形的抱枕和饰品，还有方形的开关和斑马条纹的地毯，使整个空间显得很有立体感。

260 90 05

080 80 05

095 90 30

110 92 05

260 70 05

/ Chapter 3

FAMILY-FRIENDLY

The colours demonstrate status all too clearly, alongside obvious or concealed luxury. Pure aestheticism belongs to the suboptimal attributes which knocks the socks off the clientèle or at least pulls out the rug from under their feet.

居家温情

这些色彩都十分鲜艳，华丽感时隐时现。往往那些不完美的事物才有纯正的美感，才是客人真正喜欢的东西。

If you juggle skilfully enough with these colours well, using their surface qualities appropriately as well as, the slightly dull, warmer and cooler pastel hues, then magnificent rooms for experience and recreation are bound to emerge.

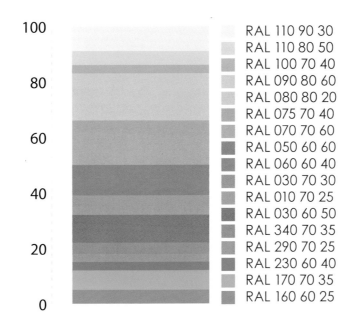

如果能熟练应用这些色彩，了解这些色彩会呈现出的效果，如时淡时浓、时暖时冷的色调，那么就能设计出愉悦身心的空间，同时还能给人一番非凡体验。

Many lightly patterned cloud pictures as well as playful-looking chessboard, dot, line, and stripe patterns were painted.

图中由点、线、条纹组成的图案，有的像形状规整的云朵，有的像趣味十足的棋盘。

FAMILY-FRIENDLY

居家温情

Those planning their hotel in a family-friendly way must achieve a balancing act between a child- and an adult-oriented environment. All the guests must feel at home, whether they are three, four, thirty-four or seventy-four.

按照"居家温情"的风格设计酒店时，应合理划分、协调儿童环境和成人环境，让所有群体都感到舒适温馨，不管是3、4岁，还是34、74岁。

TOURISTIC

爱在旅途

Cool shades are present in abundance. It seems that the bathing fun and the pool experience nurture the metaphorical content of the tourist existence.

色彩丰富、炫目，似乎暗示了游客在旅途中的沐浴、游泳之趣。

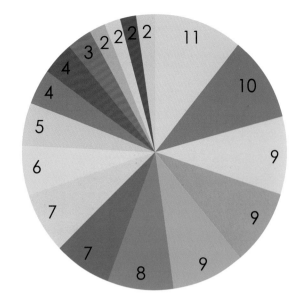

The colour combinations look like a beach towel rolled out from Agadir to Antalya or like bikinis, swimming trunks, visor caps and holiday magazines.

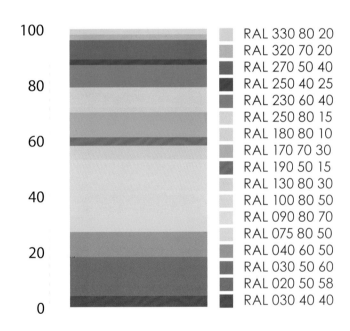

RAL 330 80 20
RAL 320 70 20
RAL 270 50 40
RAL 250 40 25
RAL 230 60 40
RAL 250 80 15
RAL 180 80 10
RAL 170 70 30
RAL 190 50 15
RAL 130 80 30
RAL 100 80 50
RAL 090 80 70
RAL 075 80 50
RAL 040 60 50
RAL 030 50 60
RAL 020 50 58
RAL 030 40 40

这些色条就像长长的彩色沙滩浴巾，从阿加迪尔铺展到安塔利亚，又如比基尼、泳裤、遮阳帽和假日杂志一般色彩艳丽。

The dense dots and circles represent the all-round recreational mood of tourism. No jags and no graphical chaos disturb the image of beautiful harmony.

密集的圆点和圆块暗指旅游中全方位的休闲气氛，既没有丝毫瑕疵，也没有丝毫混乱感，整个画面和谐美观。

TOURISTIC　　　　　　　　　　爱在旅途

The shape and colour variety is action-packed. Its playful, but also sportive hints are recorded expressively. Everyday holiday routines are interpreted as actionist models in bright colours.

图中色彩丰富、形状多样，十分有趣，还充满运动感。这些鲜亮的色彩展现了假日计划的丰富性，体现了行为主义者的作风。

/ Chapter 3

HOMELY

家常记忆

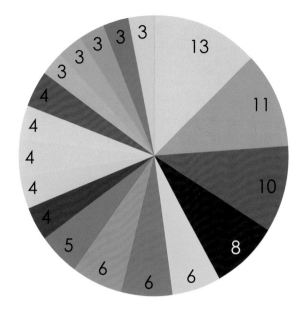

In search of colours that are reminiscent of cherry, pear or tropical woods. The colours answer the appeals for security and protect against technoid and cool materiality.

这些色彩会令人联想到樱桃、桃子和热带丛林，既不会太花哨，也不会太沉静，给人以安全感。

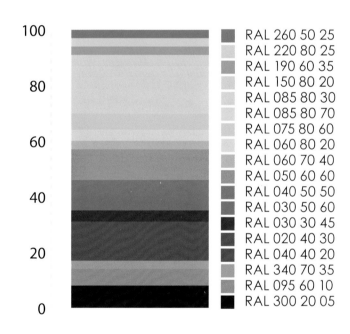

RAL 260 50 25
RAL 220 80 25
RAL 190 60 35
RAL 150 80 20
RAL 085 80 30
RAL 085 80 70
RAL 075 80 60
RAL 060 80 20
RAL 060 70 40
RAL 050 60 60
RAL 040 50 50
RAL 030 50 60
RAL 030 30 45
RAL 020 40 30
RAL 040 40 20
RAL 340 70 35
RAL 095 60 10
RAL 300 20 05

The range lists the tonal sequence of wellbeing. Here you can recognise how closely it resembles a beautiful day in autumn.

图中的色彩代表着幸福与安康，更像是美丽秋日的色彩。

Our eyes scan the homely without excitement. Not a single drawing points with its content towards agitation and dynamism.

图案平淡无奇，没有任何笔触显得凌乱，充满动感。

HOMELY　　　　　家常记忆

Does the theme contain about thirteen percent of shades of blue because bathwater in warm candlelight appears homely to us? The majority of study participants, however, felt like being in a fireplace corner with the hum of conversation and the warm and fuzzy pleasure of mulled wine or cognac.

难道是因为柔和灯光下的洗澡水令人感到温馨如家，图中蓝色调才占了约有13%的比例吗？然而，大部分研究参与者认为，图片给人的感觉像是在暖烘烘的壁炉前低声交谈，半醉半醒地饮着葡萄酒或白兰地。

/ Chapter 3

APPEALING

新奇有趣

The warm shades achieve a two-thirds majority. The appealing prevents blue from freezing solid. The clarity is convincing.

暖色调占有2/3的比例，显得十分温馨，而些许蓝色并不会给人僵硬冷淡的感觉。

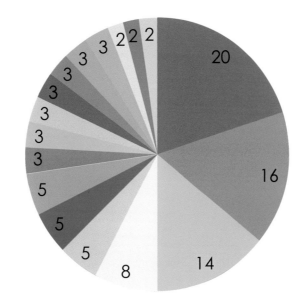

The horizontal stripes appear encouraging as well as young and playful. They contain a potential that establishes friendships.

这些色彩显得积极向上、充满青春活力、趣味十足，还隐藏着另一层含义——和谐友好。

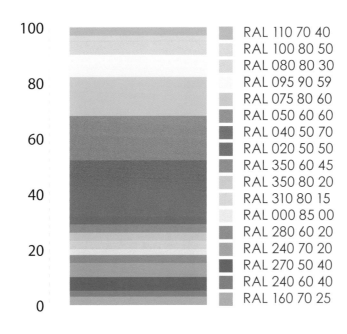

RAL 110 70 40
RAL 100 80 50
RAL 080 80 30
RAL 095 90 59
RAL 075 80 60
RAL 050 60 60
RAL 040 50 70
RAL 020 50 50
RAL 350 60 45
RAL 350 80 20
RAL 310 80 15
RAL 000 85 00
RAL 280 60 20
RAL 240 70 20
RAL 270 50 40
RAL 240 60 40
RAL 160 70 25

Round shapes with few angular elements and nothing pointed distinguish the term. We are especially attracted by what we would like to see and touch.

此图用圆形和少量带角却并不尖锐的图案来诠释"新奇有趣"，令我们对想要看到和触摸的东西充满期待。

APPEALING 新奇有趣

Simplicity and directness characterise the overall palette of colours. We have to recognise that green, brown, grey shades are not really compatible with the collective image. Each shade has a strong affinity to primary colours.

色板中的色彩与原色十分接近，整体感觉简约、直白。或许我们都能注意到，绿色、棕色和灰色与整个色盘似乎并不协调。

/ Chapter 3

FAMILY-FRIENDLY
居家温情

TOURISTIC
爱在旅途

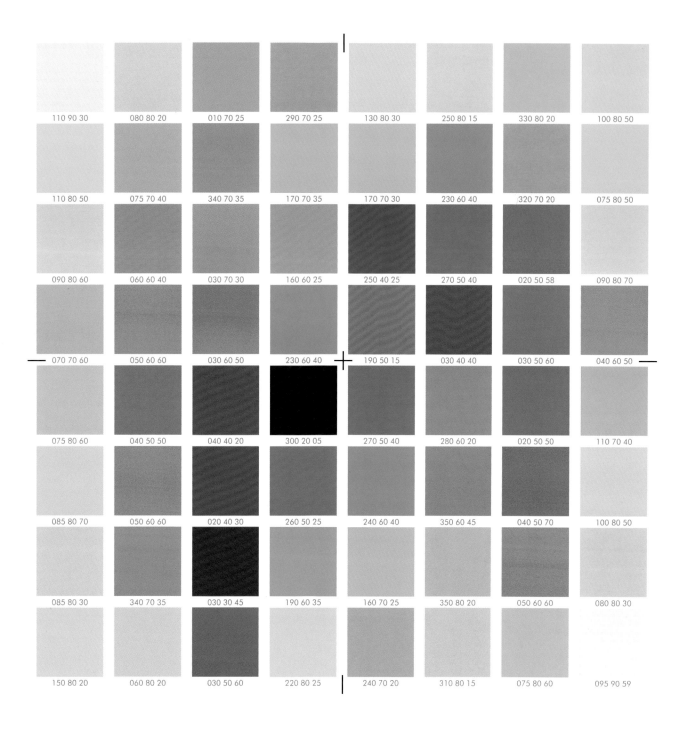

HOMELY
家常记忆

APPEALING
新奇有趣

FAMILY-FRIENDLY
TOURISTIC
HOMELY
APPEALING

居家温情
爱在旅途
家常记忆
新奇有趣

The wonderful world of holiday hotels. The colours are appealing, playful, pleasant, and they are simply fun. Even when the weather is dull or rainy, the colour scheme will ensure that low spirits do not arise.

An overwhelming prevalence of pastel hues in warm shades can be discerned. Every now and then, there are a few bright to mid-coloured shades of blue, which reflect the bright sky and the waterscape. Only very rarely will you encounter a few deep-toned accents.

假日酒店的用色丰富多彩，充满活力，迷人有趣。即使是阴雨天，人的心情也会随着这些活力四射的色彩而变得阳光起来。

整体呈暖色调，以柔和色居多，并时不时地增添些亮色和中性色彩，如暗指明朗天空和水景的蓝色，而深色则少之又少。

FAMILY- FRIENDLY 居家温情

Even if families with a large number of children are an exception, impossible to imagine the full range without suitable hotels. With increasing frequency, multi-generation families or patchwork families with an even more complicated makeup go on holiday together. There is nothing better than going on holiday to the beach, the countryside, to parks or cities and to enjoy "as a clan" a trip based on adventure, entertainment or education.

Hotel rooms should in any case feature play elements and child-oriented facilities for rest and relaxation. Internet access and game consoles should be available, of course. They are as essential as the comprehensive breakfast buffet when catering for babies and toddlers, schoolchildren and the Facebook generation of eleven-to fourteen-year olds. Cheerful, easy-care colours and robust materials make life more pleasant, instead of demanding that everyone be careful.

090 80 60   050 60 60   230 60 40   060 60 40

Generous indoor and outdoor facilities for sports and play are absolutely indispensible – unless the actual aim of the holiday was to provide for a direct encounter with pure nature and everything that happens in rivers and lakes, in mountains and valleys, in forests and fields.

The rooms should provide a high level of security and living quality and should be able to take a lot of hard wear and tear. The floors in particular must be suited to playing; they should be robust enough for romping and frolicking around and should display acoustic properties which ensure that neighbours are not disturbed. Parents, grandparents and children will be grateful and will always remember the good times they spent in the hotel.

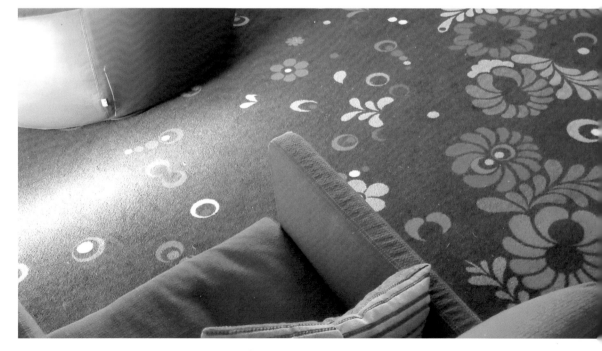

随着多代家庭和重组家庭的增多，加上各种家庭出游度假的活动也不断增多，即使孩子较多的家庭是个例外，我们也很难想象没有与之相适应的酒店是怎样的场景。到海滩、乡村、公园和都市体验一番，或者是享受一次全家游，路途中有冒险、有娱乐、有教育，还有什么比这更美好呢？

030 70 30   060 60 40   030 60 50   050 60 60   090 80 60

任何酒店空间都有必要添加娱乐设施和儿童设施，以便休息与放松。当然，互联网设施和游戏机也必不可少。同样，针对婴儿、刚学路的孩子、学龄儿童以及11至13岁的"脸书一代"，还需要食物丰富多样的自助早餐吧。喜庆、简单的色彩以及朴素、粗糙的材料能让生活更加欢愉，而严肃的色彩只会让人心弦紧绷、小心翼翼。

除了那些纯粹以接触自然，融入山河湖泊、森林田野为目的的度假方式，对于其他度假方式来说，丰富的室内外运动和休闲设施都不可或缺。

酒店客房应安全性高、居住条件好、经久耐用。尤其是地面要适合玩耍和嬉戏，而且隔音效果要好，从而不会影响隔壁住户。如果是在这样的酒店环境中度假，不管是家长还是孩子，都会欢乐无比、记忆深刻。

/ Chapter 3

RURAL

Plenty of green and turquoise, red plant shades, rich loam and poor country soil, occasional sandy shades and the yellow of ripe corn and the glow of blossoming rapeseed.

山水田园

色盘主要由大量的绿色调、蓝绿色调、红植物色，丰富的沃土色和瘠土色，以及少量的沙色、玉米黄和油菜花色组成。

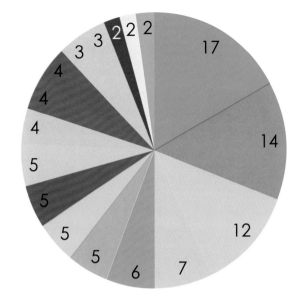

As if our gaze scans the countryside towards the horizon. Solemn and heavy, enclosed by the ground and the light blue horizon.

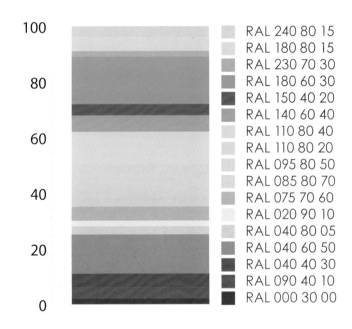

RAL 240 80 15
RAL 180 80 15
RAL 230 70 30
RAL 180 60 30
RAL 150 40 20
RAL 140 60 40
RAL 110 80 40
RAL 110 80 20
RAL 095 80 50
RAL 085 80 70
RAL 075 70 60
RAL 020 90 10
RAL 040 80 05
RAL 040 60 50
RAL 040 40 30
RAL 090 40 10
RAL 000 30 00

观看此图时就像是在平视田野，一切色彩都被大地色和浅蓝色包围着，显得庄重而肃穆。

Furrows, fields, vegetable abstractions, hills, sun, wind, and water, sluggish movement, the topography of tranquillity.

缓缓起伏的垄沟、田地和山丘，抽象的太阳和蔬菜，缓缓流淌的水流，以及此起彼伏的静谧山林，都十分生动、形象。

RURAL　　　　　　　　　　山水田园

First and foremost, this describes the earthy-vegetable original conditions, the blossoming and thriving of field crops and the changing weather from sunshine through to an overcast sky with rain clouds.

图中展现了泥土和蔬菜的最真实的一面，描绘了田野中的作物开始发芽开花的场景，更是记录了朗朗晴空忽然变得阴云密布的瞬间。

/ Chapter 3

SECLUDED

与世隔绝

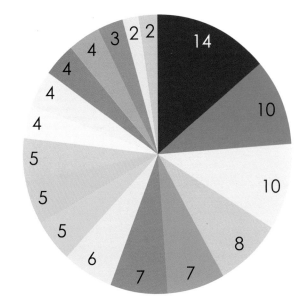

Shades without introspection; they manage without temperament. They have a logical connection which aims at objective competence.

它们是不具内在情感的色彩，不靠内在气质发挥作用。它们只是一种逻辑上的连接，关注于客观的表现能力。

Those looking for seclusion need colour schemes like this. Each shade of red, orange and violet is used to provide accents.

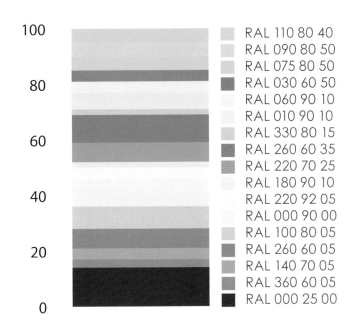

RAL 110 80 40
RAL 090 80 50
RAL 075 80 50
RAL 030 60 50
RAL 060 90 10
RAL 010 90 10
RAL 330 80 15
RAL 260 60 35
RAL 220 70 25
RAL 180 90 10
RAL 220 92 05
RAL 000 90 00
RAL 100 80 05
RAL 260 60 05
RAL 140 70 05
RAL 360 60 05
RAL 000 25 00

图中的红色调、橘黄色调和紫色调都有助于提升"与世隔绝"氛围，若想营造"与世隔绝"般的氛围，可以借助这些色彩。

Isolated circles or spherical volumes, which seem to hover independently in spaces are the graphical metaphors used most frequently.

孤立的圆圈或球形图案，似乎独自悬停于空间中。这些都是使用得最频繁的图形符号。

SECLUDED   与世隔绝

Even opposites do not occur because they would interrupt the solitude. The colours appear barren and ordinary. Here and there is a small circular area resembling a dying star in the sky.

这些色彩显得单调、普通，互相之间毫不冲突，使得这种独处状态未被打断。图中时不时地闪现出一些小圆点，暗指空中只剩幽光的星辰。

/ Chapter 3

RUSTIC

原始乡土

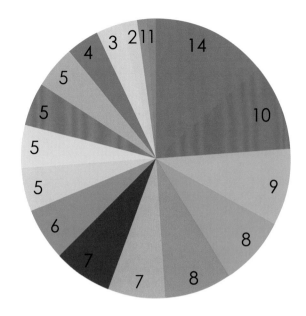

Colours of folkloristic reality, which promise security and a sturdy and safe home. Here, bed and table linen share the intense colours of the landscape.

这些色彩都是民间景象的真实反映，能给人以安全感，令人对安稳的家庭十分向往。床褥和台布往往会采用这种强烈的色彩。

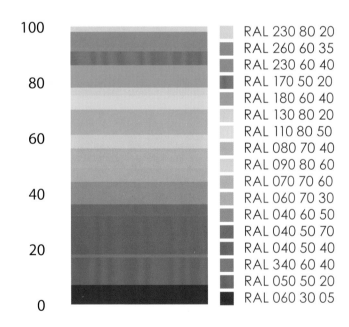

RAL 230 80 20
RAL 260 60 35
RAL 230 60 40
RAL 170 50 20
RAL 180 60 40
RAL 130 80 20
RAL 110 80 50
RAL 080 70 40
RAL 090 80 60
RAL 070 70 60
RAL 060 70 30
RAL 040 60 50
RAL 040 50 70
RAL 040 50 40
RAL 340 60 40
RAL 050 50 20
RAL 060 30 05

The colour palette is mainly earthy and autumnal-ripe. It includes medium-coloured, saturated earths, old wooden floorboards and green plants.

图中以泥土的色彩和深秋的色彩为主，如中性土壤色和深土壤色、旧木地板色和植物绿色。

We remember heaps of soil, freshly ploughed and steaming fields, heaps of earth, and seeded strips of freshly dug flowerbeds.

这让我们想起土堆、新翻耕的土地、雾气缭绕的田野以及刚播种的带状花圃。

RUSTIC 原始乡土

The rustic, rather than the rural, is the authentic expression of utilisation, tradition and work. Field-like brown areas, woody shades and meadow and forest hues contrast with occasional hints of fruit and garden.

图中以木色调、草地与森林色调以及田野的棕色调为主，并稍稍添加了一些水果色和花园色。之所以称之为"原始乡土"风格，而非"山水田园"风格，是因为前者能更贴切地反映深耕易耨、自给自足的乡村传统。

/ Chapter 3

PURISTIC

纯粹至美

Maybe that's how an engineer sees the technology of heavy machinery. The charm of puristic design is undoubtedly based on the refusal of friendly attributes.

纯粹至美的设计必然不能使用柔和亲切的色彩。或许此图恰当地描绘出了工程师眼中的科技世界——一个布满重型机械的世界。

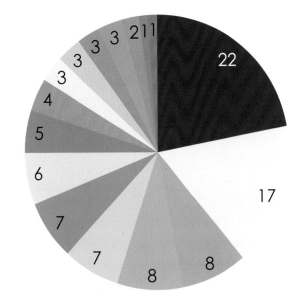

A colour range of graphic elegance. The few yellow-gold-cognac-coloured stripes emphasise the steel blue-light grey and anthracite-black bars.

这些色条组成淡雅的图形，在黄色、金色和白兰地色的衬托下，钢青色、浅灰色和煤黑色显得尤为醒目。

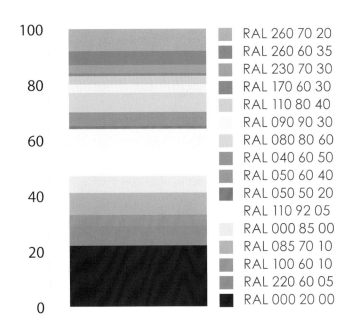

RAL 260 70 20
RAL 260 60 35
RAL 230 70 30
RAL 170 60 30
RAL 110 80 40
RAL 090 90 30
RAL 080 80 60
RAL 040 60 50
RAL 050 60 40
RAL 050 50 20
RAL 110 92 05
RAL 000 85 00
RAL 085 70 10
RAL 100 60 10
RAL 220 60 05
RAL 000 20 00

Squares, circles, diagonals, lines and intersections are the basic toys of purists.

方块、圆点、斜线和竖条都是纯粹至美的基本符号。

PURISTIC   纯粹至美

In the eyes of the study participants, purism is the constructivist part of the fundamental expectation of life. The term "rural" offers the antipode with an inimitable rural-biologi-cal quality.

在研究参与者看来，纯粹至美是生活最基本的期望之一。而"山水田园"则与之刚好相反，侧重于乡村万物的气息。

/ Chapter 3

RURAL
山水田园

SECLUDED
与世隔绝

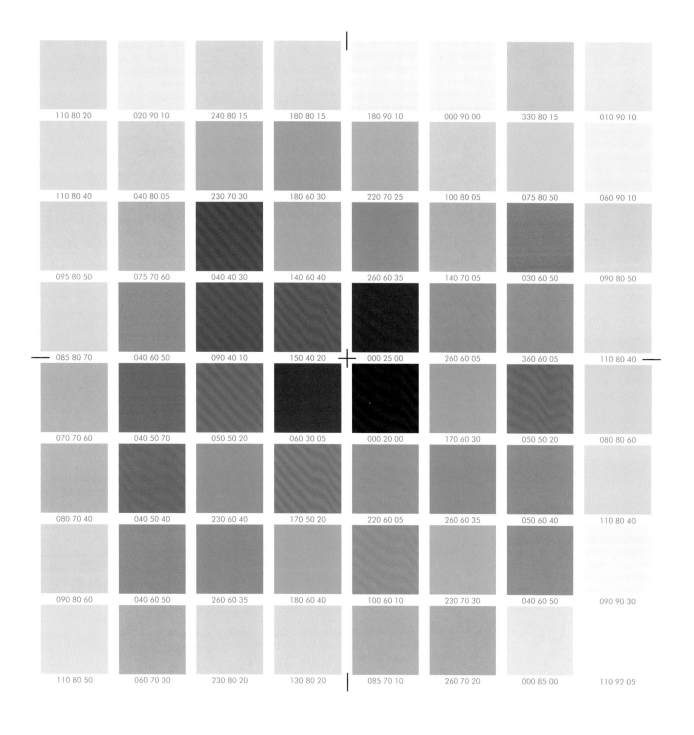

RUSTIC
原始乡土

PURISTIC
纯粹至美

RURAL
SECLUDED
RUSTIC
PURISTIC

山水田园
与世隔绝
原始乡土
纯粹至美

Perfectly contrasting qualities are combined here to form an entity. Every now and again, the desire for the ordinary, plain or rustic is overwhelming, especially for the big-city globetrotter. Then the desire for country air, fresh meadows, mown grass and sustaining food makes us head for the mountains, the sea, the village atmosphere or solitude.

The best way to escape excess, too much technology and dynamism, the permanent strain of communication is by going to where these unemphatic colours are at home.

City dwellers love the colours of the countryside because they are unaccustomed to them, and country dwellers love them because they are so close and familiar. Colours which are typical of an urban setting do not exist, except for the colours of megacities, whose scenery is characterised by steel grey and reflecting window fronts.

对比显著的色彩组合形成一个整体。人们往往更渴望回归纯朴与平凡，尤其是大都市中的环球旅行者。也正因为人们渴望乡村的空气、鲜绿的草地、割剩的短草以及基本的食物，才去登山出海，享受乡村气氛，体验独处生活。

想要远离科技和浮躁、永不停息的通信，最好的去处就是色彩浅淡各异的温馨空间。

城市居民因为好奇而喜爱乡村色彩，而农村居民是因为与乡村色彩如此亲近熟悉而喜爱它。城市没有专属于自己的色彩，当然特大都市除外，因为其风景的亮点就是钢灰色和反光的玻璃正面。

PURISTIC

纯粹至美

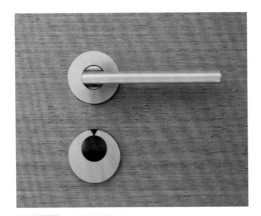

220 60 05   000 85 00

This is how we understand the puristic ambience. Factual, constructive and unadorned, which means: the matching colours and the graphic-linear style which characterises the three-dimensional, as well as the definition of the surfaces requiring structuring or decoration. In addition to grey-and-white, a bright, sunny yellow ventures to appear on the scene. Purists love to treat colours in a very vegetarian almost vegan manner, as with the subtly nuanced brighter areas and the delights that lively nature has to offer.

It cannot be denied that such design can offer great charm: from time to time, after days filled with too much excitement, it is comforting when things become too much for us and we are overwhelmed by the desire for a visual respite. When designing aesthetic meaning, simplification and omission are as important as they are for the composer, the actor and especially the cabaret artist during the interval.

It is good if there is calm without emptiness, as in the picture shown. In their stringent formalism, the silhouettes and constructivist elements that have been created incorporate invitations to relax and contemplate or to reanimate previously abandoned ideas – because no colourful curtain, bright pictures or colourful flowers disturb the creative zones of contemplation within the room. Of course, the aesthetic effects of a puristic hotel setting which is experienced with all the senses remain as a welcome option which appeals to many people especially in times when they need quiet and relaxation.

110 92 05   000 85 00   085 70 10

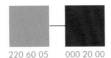
220 60 05   000 20 00

真实、积极、朴实，这就是我们对"纯粹至美"氛围的诠释。这也意味着配色和线形图案是形成立体外观的基础，说明外观也需要构建和装饰。图中不仅有黑白色的搭配，浅日光黄色的加入更是点睛之笔。纯粹主义者会以素食者或严格素食者的行为方式来对待色彩，就像图中微妙的亮色细节，显得自然活泼、愉悦人心。

000 20 00   000 85 00   080 80 60

000 85 00   100 60 10

080 80 60   110 92 05   000 85 00   050 50 20   000 20 00   100 60 10

It is the small gestures, from vases, light fittings, light switches, handle elements and a telephone to sockets at the head end of the hotel bed and the yellow colour of the small cushions, which provide proof of the experienced view of the designer. Every single element, even the smallest one, decides whether it is "well done" or only "well-intended".

220 60 05   110 92 05   000 20 00   080 80 60

000 20 00

000 85 00

不可否认，这种设计风格魅力难挡。在连续经历多个兴奋的日子之后，能有些让人想沉下心来的东西，人们反而会感到宽慰，因为此时，人们急需的是一次视觉喘息。"精"与"简"对美学价值创造的重要性，与之对作家和演员，甚至是幕间的歌舞表演艺术家同样重要。

110 92 05   090 90 30   000 20 00   000 85 00

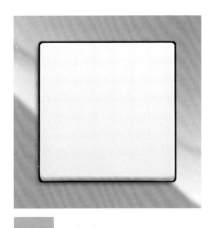

085 70 10   110 92 05

085 70 10

110 92 05

000 20 00

050 50 20

沉静而不空泛，往往效果会很好，就像图片中那样。没有彩色的窗帘、醒目的图画以及多彩的花朵，在其有力的形式中，轮廓和结构元素共同组成了适宜放松、沉思和重拾意志的创意空间。当然，对于纯粹至美的酒店布景，只有将全部感官投入，才能体会其美感，这就是其魅力所在，也是众人喜爱它的原因，尤其是那些需要宁静和放松的人们。

精致的细节往往能反映设计师的熟练度和独到见解，如花瓶、灯具、开关、手柄、床头电话和黄色小抱枕。每一个元素，甚至是最细微的元素，都能决定空间设计的成败。

110 92 05   000 20 00

/ Chapter 3

HISTORIC

历史沉淀

This range survives without any independent aesthetic substance. Principally, it shows coloured signs of old age and similar shades of an increasing contamination level of materials.

倘若将每种色彩独立开来，并不会有任何美感，而退一步看整体，这就像是岁月沉淀很久才有的色彩，就像物质受到不同程度的侵蚀而有的色彩。

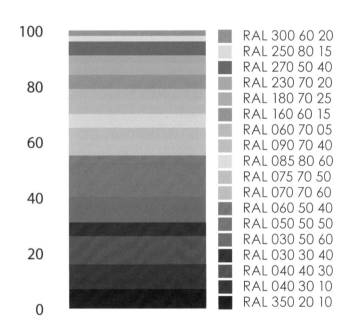

RAL 300 60 20
RAL 250 80 15
RAL 270 50 40
RAL 230 70 20
RAL 180 70 25
RAL 160 60 15
RAL 060 70 05
RAL 090 70 40
RAL 085 80 60
RAL 075 70 50
RAL 070 70 60
RAL 060 50 40
RAL 050 50 50
RAL 030 50 60
RAL 030 30 40
RAL 040 40 30
RAL 040 30 10
RAL 350 20 10

We hardly ever see pure colours. The range appears subdued and shows traces of an eventful past.

色卡中几乎看不到纯净的色彩，从而显得郁郁沉沉，似乎重现了多事过往中的种种景象。

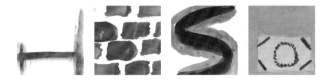

Most participants denied themselves any formal interpretation. The representations of bricks and rendered masonry are probably the most emphatic.

大多数人并不能从这些图形得出自己的见解，但都能看到砖块图案，恰似一堵墙。

HISTORIC   历史沉淀

As expected, the colours turn out to be brownish and earthy. Occasionally, the umbra, cognac and beige hues are accompanied by saturated shades of blue and green. Shadows of aging and concealment overlay the historic hues.

正如所预想的,图中以褐色和泥土色为主,还有些许暗影色、白兰地色和米黄色,并用蓝色和绿色等饱和色来增强效果。而且物体隐藏和老化形成的阴影遍布了整个"历史沉淀"色系。

/ Chapter 3

ANTIQUE

古色古香

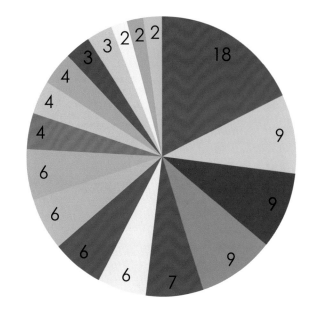

The pie chart clearly indicates the absence of warming shades and the presence of a few highly nuanced bright colours.

从饼状图中能清晰地看出，暖色调的缺失和少量明亮色调的加入。

A range that, according to expectations, banishes many things into the darkness. The darkness of the past lays a shroud over everything, especially the colours.

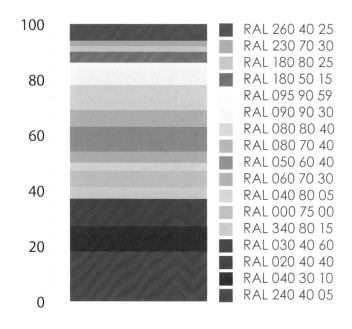

RAL 260 40 25
RAL 230 70 30
RAL 180 80 25
RAL 180 50 15
RAL 095 90 59
RAL 090 90 30
RAL 080 80 40
RAL 080 70 40
RAL 050 60 40
RAL 060 70 30
RAL 040 80 05
RAL 000 75 00
RAL 340 80 15
RAL 030 40 60
RAL 020 40 40
RAL 040 30 10
RAL 240 40 05

正如预想的那样，这是一组以暗色居多的色系，就像是过去的黑暗笼罩了一切，也影响了色彩。

Abstractions consist of old coats of paint, rotten wood, cracked masonry and charred wood.

这些图形是旧漆、朽木、碎砖和焦木的抽象形式。

ANTIQUE   古色古香

The antique is not far from the historic. The colouring of the very old is even blacker, greyer and duller than the comparative image. A few brighter values are, however, also visible.

"古色古香"色系和"历史沉淀"色系相差并不大，但前者比后者更暗、更灰、更黯淡。然而，还有些许更明亮的色彩也清晰可见。

/ Chapter 3

FEUDAL

盛世王侯

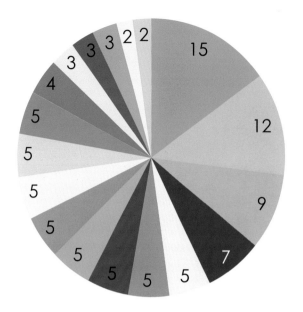

The pie chart breaks up the feudal colouring into different characteristic components. Either singly or together, the different shades show an apparently Baroque approach to storytelling.

饼状图按"盛世王侯"色系的特征将其分成不同的类别。不管是单种色彩，还是整个色系，这些缤纷的色彩明显呈现出一种巴洛克式的叙事风格。

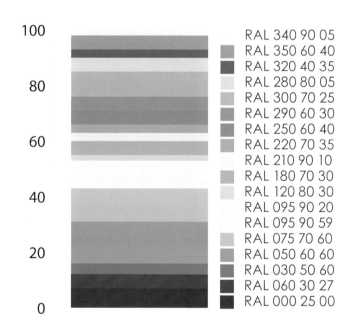

The strip can be easily imagined as the pattern of a length of fabric full of expression. Various combinations of three or four colours can form the basis of an ensemble.

这些色条很容易让人联想到图案多样、色彩丰富的长布。任何三四种色彩组合起来，都能代表整个色系。

Graphical, seemingly elegant signs are the design contents that match this theme.

这些稍显优雅的图形代表了此主题下的设计内容。

FEUDAL

盛世王侯

The number of shades of purple, lilac and violet is not surprising, and nor are the expensive gemstone and brocade shades of brass, gold, muted orange, garish green and bright and shades of dark blue.

紫红色、丁香紫和紫罗兰色频繁出现，还有丰富的宝石色和锦缎色，如青铜色、金色、暗橘色、鲜绿色、深蓝色和浅蓝色，这都不足为奇。

/ Chapter 3

PLUSH

豪华舒适

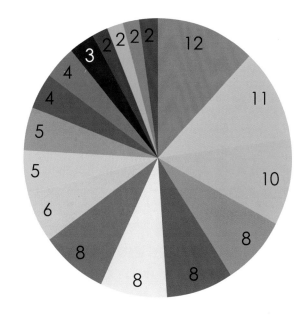

You can enjoy sitting in an armchair in these shades, regardless of whether it is with or without floral designs, scattered flowers or all-over patterns.

不管是有还是没有花纹图案，不管是只点缀着些花饰，还是满布花纹，都无关紧要，只要拥有图中的色彩，这张座椅就会令人身心愉悦。

The colour range echoes a yearning for eternity as long as plush continues to exist as a product. You will have to feel velvety yourself when sitting on and touching the furniture. Only then will you realise the extent of the feeling capacity of this wonderful fabric.

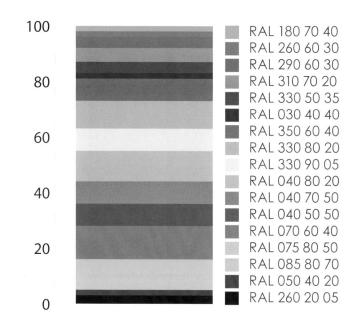

RAL 180 70 40
RAL 260 60 30
RAL 290 60 30
RAL 310 70 20
RAL 330 50 35
RAL 030 40 40
RAL 350 60 40
RAL 330 80 20
RAL 330 90 05
RAL 040 80 20
RAL 040 70 50
RAL 040 50 50
RAL 070 60 40
RAL 075 80 50
RAL 085 80 70
RAL 050 40 20
RAL 260 20 05

只要产品设计仍然坚持以舒适为标准，这组色彩就会永存。当坐在家具上，抚摸着家具，感受那种天鹅绒般的柔软时，你才会意识到这种舒适的织物对人的感觉的影响力有多大。

Here and there, roundish patterns are scattered and suggested. Formally, small drawn elements play a major role, even before the graphical interpretations.

图中到处散布着略圆或似圆的图形。从形式来看，细微的笔画最多，甚至多于能辨认出的图形。

PLUSH 豪华舒适

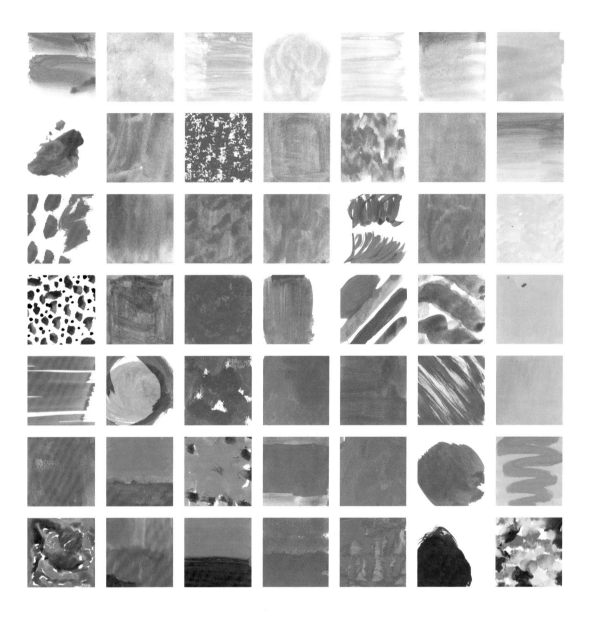

Plush is so amiable! The grandmotherly grey standard no longer applies. Biotopes of nostalgia consisting of pink, rosé, violet and lilac, are modern and kind to us all because plush is so soft, warming and soothing.

"豪华舒适"的风格会令人感到亲切无间，甚至超过了"奶奶灰"的亲切程度。由粉红色、紫罗兰色和丁香花色组成的怀旧色彩显得十分现代、亲切，甚至所有的色彩都十分柔和、温暖。

/ Chapter 3

HISTORIC
历史沉淀

ANTIQUE
古色古香

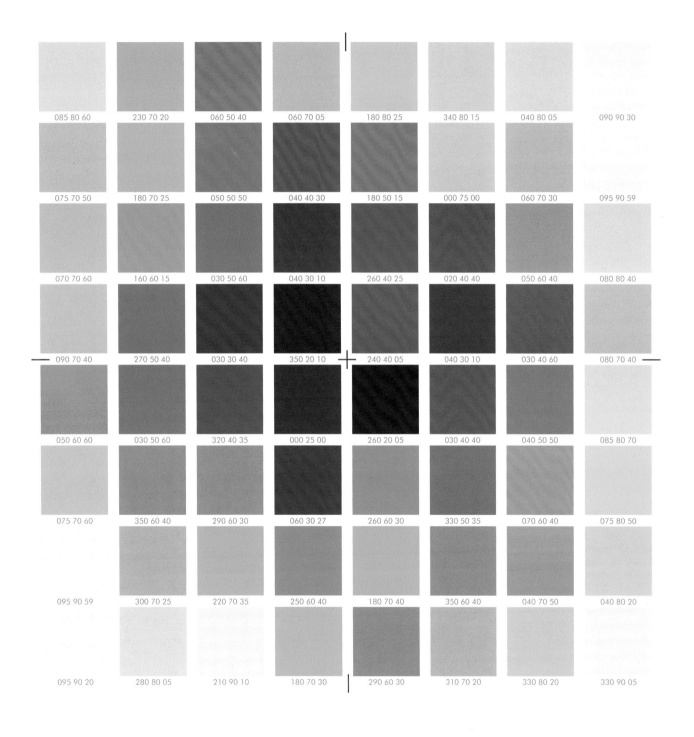

FEUDAL
盛世王侯

PLUSH
豪华舒适

HISTORIC
ANTIQUE
FEUDAL
PLUSH

历史沉淀
古色古香
盛世王侯
豪华舒适

Allowing the past to catch up with us generates a warm feeling of closeness and the desire for greater immediacy. The expectations of the past are revealed in nostalgic white lies as if the past really was more beautiful than reality, as well as more velvety, softer, more splendid and all in all more humane than the present.

If we take a look at the colour pictures which illustrate the past, we learn that the interpretations of "plush" or "feudal" when recalling the past demonstrate similar associations to "historic" and "antique".

The 64 shades on the opposite colour page feature an almost homogeneous colour spectrum in which, for example, clear blue shades are just as rare as soft shades of green. Against this prevail precious yellow and gold colours, warm red and wood hues, dark brown-black colours and prettified rosé, pink and lilac, the feudal-plush character of which is apparent. The quality of life emanating from this colour spectrum is appealing, if not incontrovertibly acceptable, but of great intensity and lively appeal.

这些色彩似乎能勾起我们的回忆，让我们感到亲切、温馨，甚至让我们想立刻捕捉到更多回忆。图中用"怀旧白"来表示回忆过往的情景，似乎过去比现实更美好、温厚、辉煌，甚至比眼前的时光都更有人文气息。

看看这些代表过去的色彩，我们便能理解"豪华舒适"和"盛世王侯"这两个主题的含义，而在回忆过去的同时，便也能对"历史沉淀"和"古色古香"产生相似的联想。

前页中的64种色彩色度均匀，十分接近。例如，图中鲜蓝色极少，因此柔和的绿色也很少。而黄色和金色、暖红色和木色、深棕色和黑色，以及玫瑰色、粉红色和紫色，则清晰地呈现了"盛世王侯"风格和"豪华舒适"风格的特征。用这组色彩设计的生活空间会很有吸引力，即使不能说是完全为人接受，至少其呈现出的活力和热情也会令人难以抗拒。

HISTORIC 历史沉淀

060 70 05 · 090 70 40 · 350 20 10 · 075 70 50 · 030 30 40

Modernism surrounds itself with borrowings from the past. Age-old lotus and palmette motifs, varied braiding, crochet, fringes and border patterns are quoted as eagerly as quilted upholstery on doors, walls and furniture. The heaps of cushions on the beds become more voluminous, the bedcovers more pompous, and the height of beds exceeds Baroque dimensions. They recall the oversize state beds in the days of Louis XIV of France, who used to negotiate important affairs of state between eiderdown and mattress and subsequently dealt with matters of hygiene by moistening his fingertips and his forehead with rosewater from a porcelain basin.

We are not fond of the realities which shed light on old times, but we like their charming stories. By the way, the origin of the short beds in those days was not that their users were small of stature, no: people sat rather than lay in their beds. Our ancestors considered this posture a safer way of not getting caught unawares by malaise or severe illness when lying down.

075 70 50 · 270 50 40 · 030 50 60

We do not preserve everything the past has to offer for the present day. The historic knowledge of the display of generosity, elegance and comfort still provides a box of tricks filled with tangible features to be adopted in order to create a well-designed ambience. Double-layered curtains in front of the windows, woollen carpets on the floor, armchairs and sofas with thick upholstery are the most intelligent, but also most tangible products to generate perfect room acoustics. The best results are achieved by fabrics with cut pile like velvet or all types of velour.

现代风尚中的许多元素都源自过去。古莲花纹、棕叶图案、各种编织样式、钩针、缘饰和边框等常用在缝制装饰织物上的手法，也用在了门、墙和家具上。床褥中堆有更多的抱枕，被套也愈加华丽，床垫的高度更是超过了巴洛克风格中的高度。这一切令人想起法国路易十四统治时期的超大号床，路易十四习惯在凫绒被和床褥上商讨国家大事，而卫生问题的解决则是用瓷盆中的玫瑰水来浸洗双手和前额。

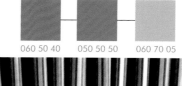

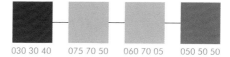

In many older hotels, ceilings, walls, columns, and passageways are restored alongside their chipped ornamental stucco. The same approach is adopted with the façades of the exterior, which are once again presented to the beholder in their old, silent grandeur. The attractiveness of historic buildings from across the ages continues to increase. Either we want to make old dreams come true, or we are simply curious about past eras, so that we can describe them to our friends. The stories that took place within these old walls and on these curved marble staircases are worthy of being added to the spirit of a hotel.

As a variation on the tale about a renowned cosmetics manufacturer who claimed that it was not lipstick he was selling but hopes, it can be said that the hotelier of a legendary hotel is not renting out rooms but mysteries. They begin with the copper-coloured light switch and are nowhere near ending at the huge, button-tufted headboard of the truly adequately dimensioned bedstead.

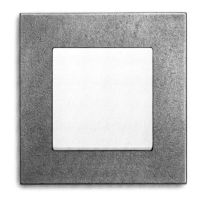

090 70 40
030 30 40
040 30 10
030 50 60

其实，我们并不喜欢旧时灯光昏暗的真实场景，只是喜欢那些引人入胜的故事。此外，古时候出现了短床褥，可这并非因为当时的人们身材矮小，而是因为当时的人们只是坐在床褥上，而非躺在其中。我们的祖先认为，相对躺着而言，坐着是一种更安全的姿势，因为这样不会突然感到不适或染上重病。

我们并不会把昨日的一切都留给今朝。然而，传统中打造大方、优雅、舒适空间的智慧仍可为今所用，因为其中给出了特征鲜明的技巧，有助于形成设计精妙的空间氛围。窗前的双层窗帘、地上的羊毛地毯、厚软的沙发和座椅，都是最具智慧、最有形的物品，有助于实现完美的感

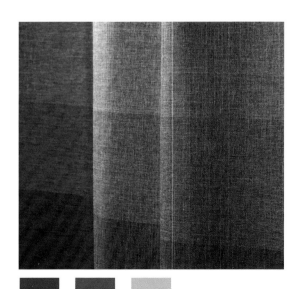

350 20 10   040 30 10   060 70 05

官体验。而天鹅割绒或各式绒毛的采用，效果则会更佳。

在许多更为陈旧的酒店中，天花、墙壁、柱子和过道的装饰会延续已有的碎裂破旧的风格。室内装饰也会采用同样的方法，从而为人们呈现古老而庄严的形象。古建筑的魅力源自年岁的积累，并且随着时间的推移不断增加。我们之所以喜爱旧建筑，或许是因为想让旧梦成真，或者仅是因为对过去感到好奇，从而有机会在朋友面前描述一番。古墙见证的故事、螺旋大理石楼梯经历的风雨，都能应用到酒店中，升华酒店的灵魂。

一位知名化妆品制造商曾说，他卖的不是口红，而是希望。如果把这个故事稍作改变，那么我们可以说，传奇酒店的老板出租的不是房间，而是神秘感。在这传奇酒店中，从铜色灯开关到大床床头板处的开关，无一不具神秘色彩。

/ Chapter 4

HOTELS BETWEEN DREAM AND POETRY: IDYLLIC TO NOSTALGIC

When travelling by ship on the river Rhine, starting from Cologne and heading upstream past Koblenz, you will arrive at where this chapter heading is leading us to. The dreamlike is very close to the poetic. A highlight is the Loreley, and as we pass, the yearning, seductive sounds will inevitably come to mind. Castles and hotels, mansions and villas located on the port and less frequently on the starboard side drift by. Similar memories arise when travelling on Lake Lucerne from Lucerne to Weggis and Vitznau, when looking at the imaginative, inspiring hotel façades which positively sparkle, catching people's attention with the rippling water reflected on the surfaces of the windows.

If the temptations announced on the outside are confirmed on entering and our expectations are largely exceeded, then our enthusiasm and amazement will be great. On the next eighty pages, we shall take six examples to show what the hotel life of the sophisticated connoisseurs may look like.

Soulful atmospheres alternate with fanciful, glamorous or fabulous and creatively whimsical ones. The creative scope opened up by simply mentioning the terms is vast and exhilarating, just as a bystander watching someone bite into a lemon will instinctively purse his lips and suck in his cheeks.

When looking at the staff of such hotels, we notice that a special charm seems to determine their physiognomies; their gentle smiles, their friendly eyes, their expressions and gestures, the very choreography of the way they move and stand.

In recognising it, we also speak of the "special spirit" of a hotel, which makes itself felt throughout the building. The fact that it actually exists is proved true more often than we think. If we experience affection, it has also to do with our behaviour. As a guest we take upon ourselves the obligation of becoming part of a poetic mise-en-scène whenever we stay in an idyllically located and furnished hotel.

Incidentally, being a pleasant guest is fun and makes us happy without taxing us unduly. You can enjoy the amenities of hotel and service, of kitchen and bar even more. The fundamental condition for this is an ambitious design of all rooms which has been accomplished in a people-friendly way and to a high standard.

梦幻诗意酒店：
从诗情画意到温婉怀旧

从科隆出发，泛舟从莱茵河逆流而上，途径科布伦茨，便能到达本章将带领我们去向的地方。梦幻风格与诗意风格十分相近。沿途所经过之处，不得不提的就是罗蕾莱，这里的思念之歌扣人心弦，令愁思涌上心头。城堡和酒店，豪宅和别墅，都矗立于港口，右舷河岸上则很少出现。同样，从卢塞恩坐船穿过卢塞恩湖到维茨瑙，看到极富想象力的酒店外观，人们的脑海中也会浮现相似的记忆，荡漾的水波闪闪发光，映在酒店的窗户上，令人神往。

如果酒店极富魅力的外观能吸引我们进入，而且进入之后所感受到的魅力更是大大超过了我们的预期，那么我们会热情高涨，惊讶万分。接下来，我们将用6个例子来展示高级鉴赏家的酒店生活是怎样的。

想必这类酒店逸韵高致、奇特迷人且变化多端。当听到这些词语，人的脑海中就会不断浮现各种创意想法，内心会激动无比，就像看到别人咬开柠檬，自己也会不自觉地噘着嘴吮吸。

观察此类酒店中的职员，我们发现，似乎有种魔力使他们的外貌都英俊美丽，笑容都灿烂甜美，眼神友善真诚，言谈轻柔，举止文雅。不管是站姿，还是走姿，都像舞姿那般曼妙轻柔。

酒店的特色精神能贯穿整栋建筑，其真实的力量比我们想象的要大得多，为了识别它，我们也对此进行了介绍。喜爱之情的产生往往也与行为相关。不管是在简陋朴素的酒店中，还是在家具齐全的酒店中，作为客人，我们往往会尽力让自己成为诗意情景中的一部分。

偶尔做做客人，负担也并不繁重，还会令我们觉得快乐、有趣。酒店的设施与服务、厨房和酒吧也会令人觉得十分享受。然而，这些都是建立在完善、高档、宜人的设计之上的。

238	idyllic - romantic - bewitching - enchanted	诗情画意 – 浪漫多情 – 妖媚娇艳 – 魔法情缘
250	festive - uplifting - dreamy - poetic	节日颂歌 – 引人向上 – 如梦如幻 – 朦胧诗意
264	lively - creative - imaginative - impressive	活泼灵动 – 天马行空 – 想入非非 – 惹人注目
276	inspiring - motivating - entertaining - tempting	心驰神往 – 激发斗志 – 娱乐至上 – 鲜艳夺目
288	multi-cultural - international - classic - folkloristic	多元文化 – 国际风范 – 经典传承 – 地域风俗
304	fairytale-like - cosy - snug - nostalgic	恍如童话 – 亲密无间 – 舒适温暖 – 温婉怀旧

IDYLLIC

Nature is light and fresh, overlaid with dew and frost. The search for the idyllic is the tragicomic yearning for paradise.

诗情画意

大自然色明亮、清新，有时被露水洗礼，有时被白霜裹住。寻找田园般的去处，无异于寻找乐园，是一个悲喜交加的过程。

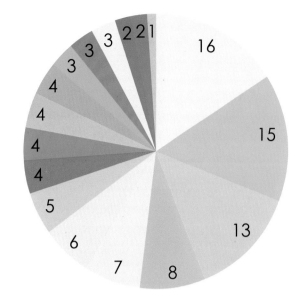

The stripes are sorbet-coloured and finished with a milky coating. Their natural origin cannot be denied.

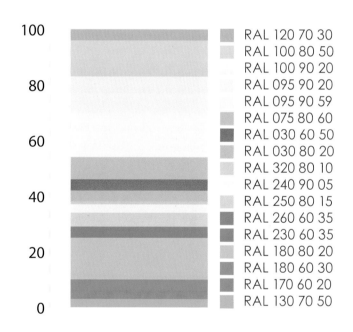

RAL 120 70 30
RAL 100 80 50
RAL 100 90 20
RAL 095 90 20
RAL 095 90 59
RAL 075 80 60
RAL 030 60 50
RAL 030 80 20
RAL 320 80 10
RAL 240 90 05
RAL 250 80 15
RAL 260 60 35
RAL 230 60 35
RAL 180 80 20
RAL 180 60 30
RAL 170 60 20
RAL 130 70 50

冰糕般的色彩，似乎又在牛奶中浸过，蒙着一层乳白外衣。不可否认，这些色彩都源自大自然。

Hints of grasses, flowering plants and grasslands. And we can recognise watery areas like ponds and marshy meadows as well as the evanescent.

这些图案就像是小草、草地和开着花的植物，又像是水池、沼泽草地和凋零的花朵。

IDYLLIC 诗情画意

Colours as if illustrated with a panpipe. Green alpine pastures and clear mountain lakes and the endless blue sky above. Romantic idealisation cannot exist without dream-like, airy colour landscapes.

这些色彩就像排箫一般,纵横有序,描绘出了翠绿的高山牧原、清澈的山中湖泊、无穷尽的蓝色天空。浪漫的想象往往离不开色彩柔美的、梦幻般的风景。

ROMANTIC

浪漫多情

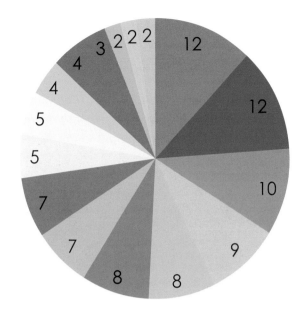

What is a bunch of pink-red roses compared with this colour circle for the friendship book? Do not look at it for too long at a time! Just a single leftover dab of colour makes it even more beautiful.

如果把色盘比作粉色和红色的玫瑰花束，是否恰到好处呢？记得一次别看太久！淡淡的一抹色彩，更加增添了美感。

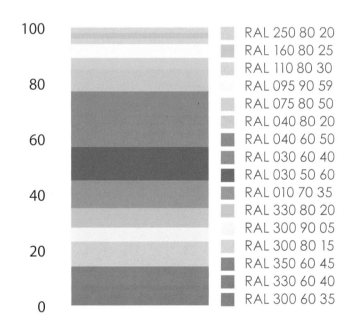

The colour stripes display an attractive tension. The light blue and the concluding green-turquoise shades reflect a lively watery depth.

色彩条颇具张力，十分醒目。浅蓝色和蓝绿色并列起来，就像是水的深度在变化。

Hearts are essential. Fruit, flower-like patterns, rainbows, shining sections of sky ensure a romantic centre.

在"浪漫"主题下，心形图案是基本要素，而水果、花朵、彩虹和闪亮的星空，都是营造浪漫气氛的关键。

ROMANTIC 浪漫多情

Violets, rose-red and lilac-like hues represent the visual spectacle of the bright red enchantment of souls and the beautiful appearance of the adamant, the everlasting and the rarely non-sentimentalised.

紫色、玫瑰红和丁香花色组合起来，从视觉上展现了鲜红的心的魅力，而美丽的外观之下，又隐藏着坚定、永恒和感伤。

/ Chapter 4

BEWITCHING

妖媚娇艳

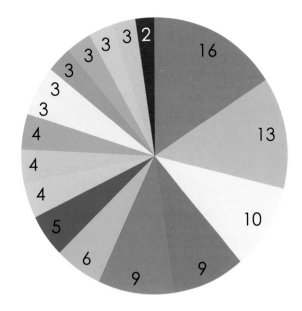

Grey and black are missing in the magicians' colour range. The mystic, bluish tinged shades of red are the favourite colours of illusionists.

灰色和黑色从魔术师的色盘中缺失了，而神秘的、微蓝的浅红色调成为了他们最喜爱的色彩。

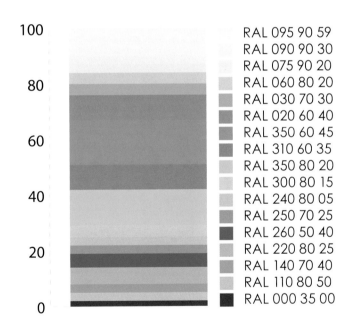

RAL 095 90 59
RAL 090 90 30
RAL 075 90 20
RAL 060 80 20
RAL 030 70 30
RAL 020 60 40
RAL 350 60 45
RAL 310 60 35
RAL 350 80 20
RAL 300 80 15
RAL 240 80 05
RAL 250 70 25
RAL 260 50 40
RAL 220 80 25
RAL 140 70 40
RAL 110 80 50
RAL 000 35 00

The colours seem hauntingly grave and not at all playful. They bear medieval traces of seduction and sedition.

这些色彩显得黯然销魂，不带一丝玩味，似乎保留了处于诬惑与霍乱之中的中世纪的痕迹。

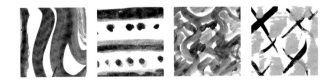

The serpentine lines and patterns squirm and writhe in various drawing styles.

蛇形的线条和图案似乎在蠕动、在翻滚，最终形成风格不同的图画。

BEWITCHING　　　　　　　　　　妖媚娇艳

Magicians have always had a predilection for bewitching lilac. Blue and red cloths are also extremely popular tools of the nimble-fingered and -spirited apprentices and their masters.

魔术师往往偏爱迷人的淡紫色。对于心灵手巧的魔术学徒和大师来说，蓝色和红色布条是特别受欢迎的道具。

ENCHANTED

魔法情缘

The circle of colours looks noble. It reminds us of the bygone age of chivalry with fairies and fanatics at royal courts. Staying the night in an enchanted castle is everybody's spooky wish.

这些色彩有种高贵的特质，让人不禁联想到逝去的骑士时代，以及分为绅士和狂徒的皇庭。夜晚在魔幻城堡中逗留，是每个人的噩梦，令人毛骨悚然。

Even though the shades are lined up in a well-ordered fashion, they cannot deny their darkly complex mystical origin.

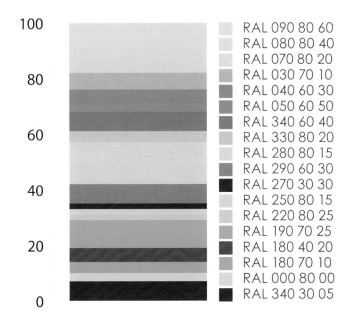

虽然各彩色条排列有序，但仍然无法隐藏其黑暗、复杂而神秘的源头。

The signs are clear: they range from protecting crosses to rune-like zigzag patterns through to innocent, bewildering spirals.

这些图案指向清晰：防护架、符文般的锯齿图案，以及简单而又令人眼花缭乱的螺旋图案。

ENCHANTED 魔法情缘

This is how the study participants imagine the enchanted castle: cold blue and black crosses, blood-red, blue silk and white-grey mould-covered areas hidden behind caput mortuum.

冰冷的蓝黑十字架，以及血红色、青丝色和白灰色区，隐藏在骷髅之下，这就是研究参与者想象中的魔幻城堡。

IDYLLIC
诗情画意

ROMANTIC
浪漫多情

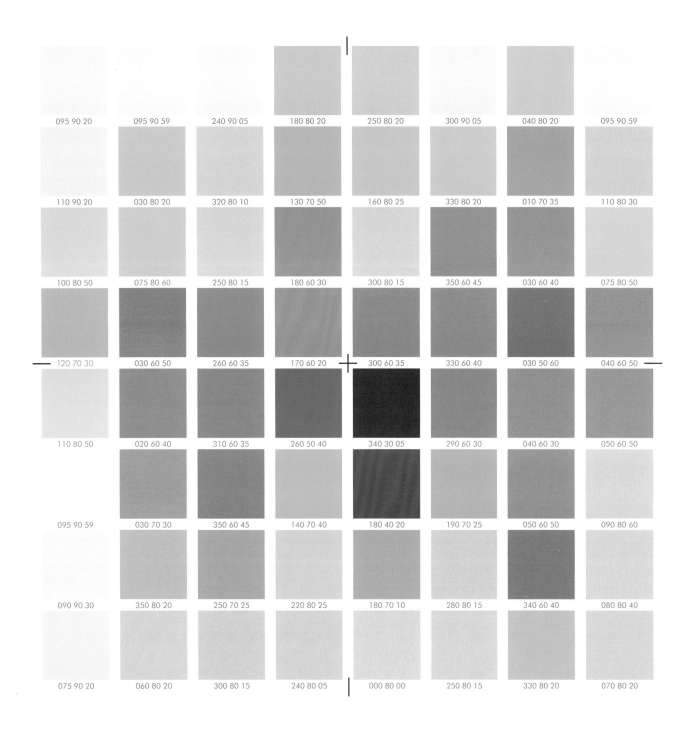

BEWITCHING
妖媚娇艳

ENCHANTED
魔法情缘

IDYLLIC
ROMANTIC
BEWITCHING
ENCHANTED

诗情画意
浪漫多情
妖媚娇艳
魔法情缘

The basis of the colour ensemble is to be found in the light cream and sorbet colours, the water and pond shades and the vernal blossom hues. Dark colours only appear twice. Friendly colours are invariably assigned positive connotation. The shades used to represent an amiable atmosphere are always bright, fresh and friendly.

Dark brown is absolutely incompatible with these dancing colours. Two vibrant dark shades exist solely with the adjective "enchanted", where the mystical and mysterious, but also large amounts of the fairytale atmosphere will be in demand. It is perfect if a hotel has a soul; in modern English this will be described as the USP – the Unique Selling Proposition. Especially in a world of functions and objectivity and competencies, of mathematical formulas and frequent appeals to common sense, the emotional argumentation always remains the more powerful one. Neuroscientists explain why: feelings are older and thus more reliable than thinking. And that's why every now and then one should rely on intuition rather than cognition.

整个色系是以浅奶油色调、冰糕色调、水塘色调和鲜艳的花色为基础的，其中仅有两处出现了暗色。鲜亮、柔和的色彩往往会被赋予积极的含义，因此也常被用来营造温馨的氛围。

显然，深棕色与这些轻快的色彩并不兼容。独立的两块深色区域契合了"魔法情缘"主题下的神秘感，但大量的童话般的色彩也不可或缺。完美的酒店应拥有自己的灵魂，用现代英语来讲就是USP——独特卖点。尤其是在注重实用性、客观性和竞争力，注重数学公式和效率的世界，情感上的说服力将更加重要、更加有力。神经学家为此提供了解释：情感发展比思维发展更长久，因此更加可靠。这就是为何要多听从直觉而非认知的原因。

ROMANTIC 浪漫多情

A purple-coloured velvet curtain partly screens the restaurant corner, which is immersed in a subdued, warm light. A romantic mood does not necessarily require radiant sunlight. We feel more at ease in a relatively calming, subdued and yet partly stimulating atmosphere. Besides "modernism", "country style" and "avant-garde", the "romantic" is an evergreen as regards furnishing concepts, and is also a recipe for a better life. It has long been established across large parts of Europe. From time to time, predilections vary according to the spirit of the times. Thus the passion for the countryside continues unabated, and modernism is en vogue as a meeting place of the not yet decided. The avant-garde is overtaken by the romantic.

350 60 45   300 90 05

We regard the formal contents, which extending right into the playful, as equally compelling and graceful and amiable. The furnishings have a feminine touch and mainly feature arched, gently curving lines. Decorative elements play a more important role than, for example, in function-oriented modernism. The examples indicate how colours and shapes and surface feelings complement each other and create a truly intentional, historicising design picture. The design references range from elegant and chic to bright red manifestations and snug closeness.

紫色的天鹅绒窗帘遮住了酒店的一角，又被柔和的灯光笼罩。浪漫的气氛并不一定要灿烂的阳光，因为相对平静、柔和且带些许醒目色彩的空间会更令人感到舒适自然。除了现代、乡村和前卫等风格，浪漫风格作为装饰理念将盛行不衰，同时还是打造美好生活的秘方。其实，这种风格早在欧洲各地盛行。虽然人们的偏好会随着时代精神的改变而改变，但人们对乡村风格的热情却有增无减，现代与时尚之间的界限仍无法划清。但前卫风格却被浪漫风格取代了。

040 60 50   300 60 35   300 80 15

"形式"直接发展到有趣，进而变得惹人注目、优雅亲切。这些家具有着弯曲的线条和圆润的外观，充满女性气质。相对注重实用性的现代风格而言，装饰元素在浪漫风格中发挥着更重要的作用。所列图片说明了色彩、形状和外观触感三者是如何达到互补，形成一张精心设计的、有历史感的设计图的。可参考的设计风格包括优雅、别致、鲜红热烈和舒适温馨。

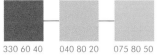

330 60 40   040 80 20   075 80 50

FESTIVE

节日颂歌

The colour circle indicates a collective agreement as to the nature of festivity: lively, exhilarating, accompanied by a few dashes of lightness and black attitude.

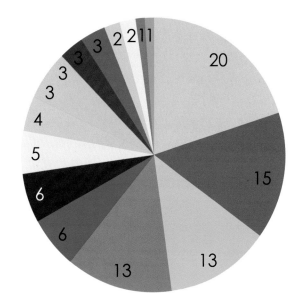

研究参与者集体认为色盘中的色彩体现了"节日颂歌"的实质——热烈欢快，同时也夹杂了些许浅色调和深色调。

A range that is not exactly filled with tension! It is without conflicts and is suitable for all generations. It is also quite colourful and scarcely tainted with references of status.

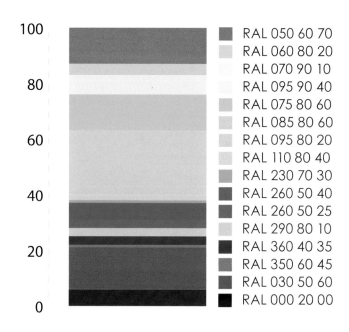

RAL 050 60 70
RAL 060 80 20
RAL 070 90 10
RAL 095 90 40
RAL 075 80 60
RAL 085 80 60
RAL 095 80 20
RAL 110 80 40
RAL 230 70 30
RAL 260 50 40
RAL 260 50 25
RAL 290 80 10
RAL 360 40 35
RAL 350 60 45
RAL 030 50 60
RAL 000 20 00

这组色彩不含张力，也没有任何冲突，适合所有人群。而且色彩都十分鲜艳，也未曾被用作某种身份的代表色。

Dynamics dominate the drawing style: serpentine, diagonal and wavy lines.

蛇形线条、对角线以及波浪线，都极富动感，成为图案的主导风格。

FESTIVE 节日颂歌

That's what "festive" looks like: plenty of red and gold, with a portion of blue and a few sorbet shades as additional ingredients. Pageantry is kept within reasonable limits. The focus of the interpreters points towards feasting, fun and atmosphere.

大量的红色和金色以及充当点缀的部分蓝色和些许冰糕色，恰到好处地描绘了节日的场景，华丽而不过分奢华。这种阐释是针对宴会、娱乐和节日气氛的。

/ Chapter 4

UPLIFTING

引人向上

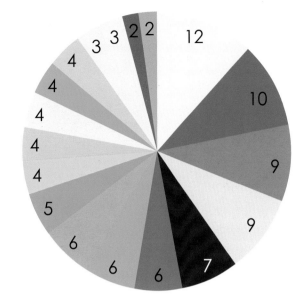

The amounts of colours, regardless of whether they are cold or warm, balance each other. Their principal claim firstly aims at the precious and then at power.

色盘中的色彩先是缤纷夺目，进而又很有力量感，但不管是冷色还是暖色，都能互相平衡。

The range carries a meaning. The shades are those that were used for Renaissance paintings, plus the patina of age which darkened them. Over the years, what was originally powerful staggers towards the tragic-dramatic.

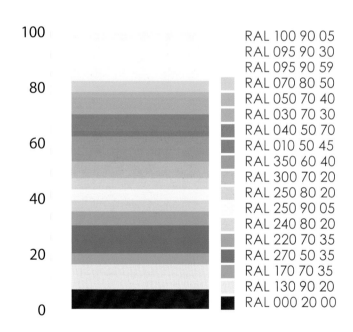

RAL 100 90 05
RAL 095 90 30
RAL 095 90 59
RAL 070 80 50
RAL 050 70 40
RAL 030 70 30
RAL 040 50 70
RAL 010 50 45
RAL 350 60 40
RAL 300 70 20
RAL 250 80 20
RAL 250 90 05
RAL 240 80 20
RAL 220 70 35
RAL 270 50 35
RAL 170 70 35
RAL 130 90 20
RAL 000 20 00

这些色彩具有某种含义，也是文艺复兴时期绘画采用的色彩，因该时代的特征而显得暗淡。历史中，曾经权如泰山撼而不动的，最终都在风中摇曳，化成悲剧。

Many signs thrust upwards. The diagonal lines and bars are more important than the vertical pillars.

图中上凸的画痕较多，而且对角线和条纹比竖向线条更加凸出。

UPLIFTING　　　　　引人向上

A colour scale with an intellectual, aesthetic touch. This is revealed by amounts of blue and black. The ecclesiastical purple emphasises the appeal of the sovereign claim.

图中以蓝色和黑色为焦点，使得整组色彩颇具美感和理性。而为教会所用的紫色突显了权力的诱惑力。

/ Chapter 4

DREAMY

如梦如幻

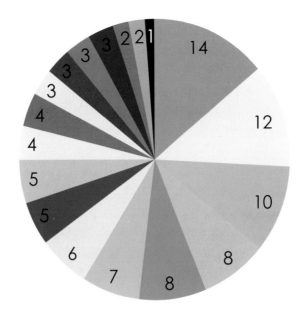

The colour variety is astounding. Surveys indicate that most people dream without specific colour memories. Some, however, dream vast colour images and turbulently colourful adventures.

颜色种类之多令人惊叹。调查结果表明，大多数人的梦境都与具体的色彩记忆无关。然而，有些人会梦到大量彩色画面，以及多彩而激烈的冒险。

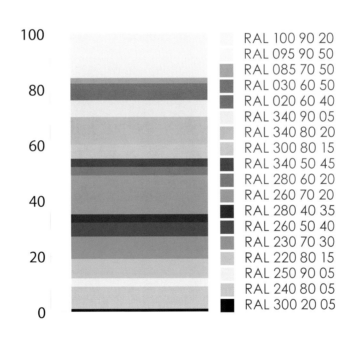

RAL 100 90 20
RAL 095 90 50
RAL 085 70 50
RAL 030 60 50
RAL 020 60 40
RAL 340 90 05
RAL 340 80 20
RAL 300 80 15
RAL 340 50 45
RAL 280 60 20
RAL 260 70 20
RAL 280 40 35
RAL 260 50 40
RAL 230 70 30
RAL 220 80 15
RAL 250 90 05
RAL 240 80 05
RAL 300 20 05

The dream of a hotel can also look like this. Why not? It is good that the shades white and grey are dulled or brightened up respectively.

梦中的酒店也可以有这些色彩，有何不可呢？分别将白、灰色调暗化或亮化，效果也会很好。

In addition to two-dimensional paintings that are mostly lightly illustrated, there are also large-scale, motionless statements.

这些平面图案大多是轻描淡画出来的，显得丰富饱满、安然静止。

DREAMY　　　如梦如幻

Dreams seem to be coloured either light blue or rosé. The contents are rarely concrete, but are frequently blurred and pallid. There is a preference for amorphous surface paintings and plain-coloured pictures.

梦的色彩似乎多是浅蓝色或玫瑰色的，其内容很难十分具体，时常模糊黯淡，就像这幅图案变化不定、色彩平淡无奇的画。

/ Chapter 4

POETIC

Clear pastel hues in harmony with to nature resemble meadows with small creeks, lake colours and golden-brown crop shades.

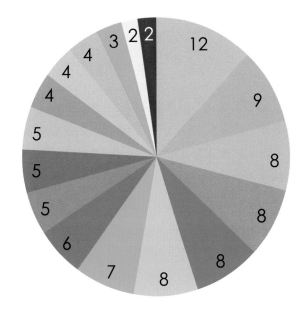

朦胧诗意

清晰、柔和的色彩仿佛源自大自然,就像是草地、小溪和湖水的色彩,还有谷物的金褐色。

Frequently, the colour stripes are applied in medium shades. They reflect nature with reserve and a cool sense.

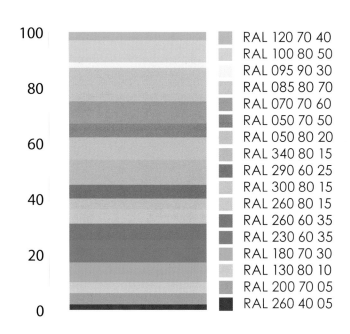

图中许多都是中性色彩,反映出保守、冷静的自然特征。

Many drawing contents with interpretations close to nature: animals, park scenes, fields of fruit, forests, bushes, and geological structures.

画中的图案很接近自然,如动物、公园、果园、森林、灌木丛以及地质构造。

POETIC 　　　　　　　　　　　　　　　　朦胧诗意

Colours as they can only be generated by a spring poem. We recognise a roundelay full of plain simplicity rather than a dramatic ballade with charming catastrophes.

这些色彩只有在颂春诗中才会出现，从中似乎能读出旋律简单的回旋曲，而非拥有悲壮故事情节的叙事曲。

/ Chapter 4

FESTIVE
节日颂歌

UPLIFTING
引人向上

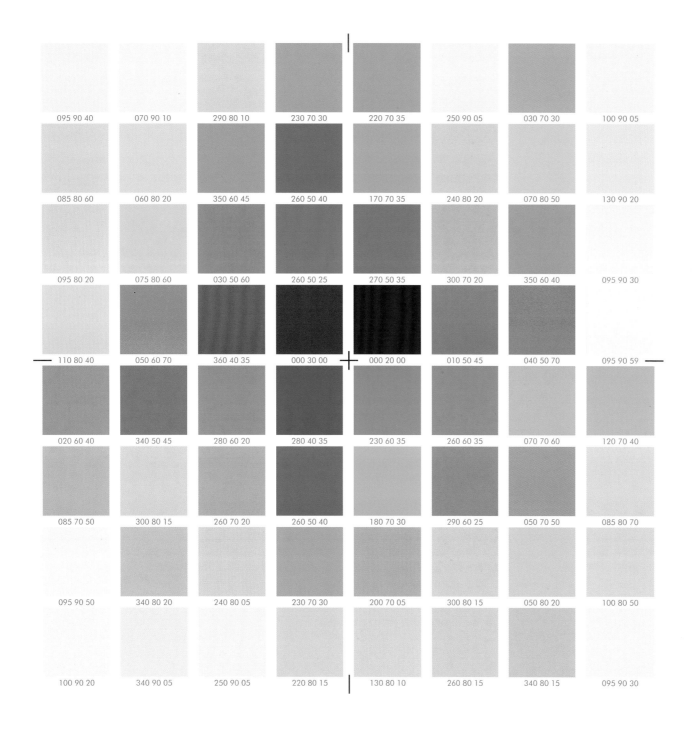

DREAMY
如梦如幻

POETIC
朦胧诗意

FESTIVE
UPLIFTING
DREAMY
POETIC

节日颂歌
引人向上
如梦如幻
朦胧诗意

The colour range consists of a generous range of blues, a good-natured floral one and a naturally coloured yellow-green spectrum. Every single colour on its own and all of them together evoke atmospheric pictures and emotional templates in our mind's eye. They are not strange to us and are frequently related to events we have experienced or which are taking place in our imagination. Here, reality and dream are brought together. They create the basis for the development of a colour-image capacity which acts as a harmonious complement and is thus able to create programmatic models.

Use the creative possibilities the template offers. The most interesting, light-hearted or vigorous or purely aesthetic combinations, imbued with practicable, profound or functional characteristics will arise by coincidence or through experimental control. In design, too, we should pay more attention to the principles of experiment, coincidence or chaos. The control of systematic orientation is not just matter of awareness, but it is equally successful and equally strategically effective.

这组色彩主要由大量蓝色调、柔和的花色和自然的黄绿色调构成，但不管是单种色彩，还是整组色彩，都能让人眼前浮现气氛独特的画面和饱含情感的场景。这些对我们来说并不陌生，而且时常出现在我们所经历过的或者所想象过的情境中。现实和梦境在此融为一体，为色彩想象能力的发展提供基础，这种能力既是一种和谐的互补，又能创造出规划有序的范例。

充分发挥这组配色方案提供的可创新性。不管是最有趣或是最轻松，还是最活泼或是仅存美感的色彩组合，都应与深刻性、实用性、可行性相结合，这些特征会在巧合或实验控制中出现。同样，我们在设计时更应关注实验原则、巧合或混乱的情形。整体效果的控制不只是意识的问题，更是成果和战略上的成效。

POETIC 朦胧诗意

340 80 15   095 90 30   230 60 35   050 80 20

Perhaps it may not signify the most lucrative of businesses, but it is definitely a sure marketing tool: a hotel with an interior like this and the corresponding accessories calls for its own library and a "book boutique" or "librairie" devoted to lyric poetry, or a "book shop" or a "Bücherladen". Regardless of what we may choose to call it.

The value demands of existence change dramatically: instead of the search for meaning by psychology, we find philosophy. The number of philosophical practices is growing. The assessment leads away from a beautiful and glamorous appearance and towards philanthropic, i.e. humane conduct. Culture and science, architecture, art, and poetry are the sustainable characteristics of the present and the ideal scale for the future. By the way, hotel marketing has always been years ahead of the times: think of the terms globalisation and internationality; think of the need for fun and adventure, for wellness, sporting and fitness; think of castles and palaces, country houses and farmhouses and lodges, of romantic idylls and poetry. The hotel industry had and still has a good nose for sensing what the next trend will be.

330 80 20   300 90 05

100 80 50   290 60 25   095 90 30

The "poetic hotel" also calls for narrative contents. In this respect, the name plays a major role, the façade as well as the location. Here, the good winemaker brands lead the way. The design must represent the thematic orientation in the smallest details. That the rooms are equipped with pictorial patterns applies to wallpapers, fabrics for curtains, upholstery, bedspreads, and cushions as well as the floor. Lamps, light fittings, large and small bowls, jugs and containers of all sorts, pictures including picture frames must also support the great overall design with accents. When it comes to the fine details of the decor, the smaller objects in particular attain a special significance in contrast to either size and scale. Flower patterns are en vogue. Today, they are drawn fairly realistically and frequently finished with satin, shimmering and opalescent surfaces. Patterned materials have an advantage compared with plain-coloured surfaces. They have highly stimulating characteristics; they call upon people to give an opinion. Patterns never leave people indifferent but prompt them into action.

180 70 30   260 80 15   260 60 35

260 80 15   130 80 10   200 70 05

POETIC

有此般室内空间的酒店，还有专门收藏抒情诗类书籍的图书室，你也可以称之为精品书吧，或者小书吧，抑或是书店。也许这并不是最有利可图的交易，但必定是有效的营销渠道。

如今，生存的价值需求转变巨大，从寻求心理需求转向了信仰上的需求，而且信仰方面的尝试也不断增多。评价也从外观是否美丽动人转向了内在是否善良友爱，如人道行为。文化、科学、建筑、艺术和诗歌等方面，都是衡量当今社会的可持续发展程度的指标，也是衡量未来社会的理想程度的指标。同时，酒店营销一直具有超前性：从全球和国际的角度来思考语言；从趣味、冒险、健康、运动和健身等方面考虑需求；还有风格和种类方面的思考，如城堡、豪宅、乡间住宅、农舍和旅馆等种类，以及浪漫、田园和诗意等风格。酒店行业始终对潮流十分敏感，总能把握未来潮流的走向。

诗意风范的酒店也应有叙事内容。从这个角度来讲，命名是关键，外观和位置也不可小视。就如良好的酿酒师形象总是能起领头作用那样，设计中最微小的细节也应与主题方向一致，如墙纸、窗帘、座

095 90 30　　200 70 05

200 70 05　095 90 30　050 80 20　260 40 05

130 80 10

095 90 30

200 70 05

050 80 20

260 40 05

套、床品、坐垫以及地面的图案，灯具及其饰件，或大或小的碗、壶和各类容器，带有画框的画作，都应具有提升整体设计格调的功能。当涉及装饰细节时，不论大小还是规模，微小的物品能发挥更大的作用。花式则应新颖时尚。而今，逼真的图案、光滑的表面以及亮乳白外观更为流行。有图案的材料也比单调素色表面更胜一筹。那些激动人心的特征往往能吸引人们来评价。图案从来不会给人以冷漠的回应，只会迅速唤起人们的行动。

/ Chapter 4

LIVELY

How wonderful that a lonely linden green and a few splashes of black are hidden here. These are shades that earn our respect as a minority opinion and thus enjoy protection as exotic species.

活泼灵动

些许椴绿色和黑色隐藏在色盘中，显得如此奇妙。这两种色彩虽然所占比例很少，却也因此倍显珍贵，就像外来物种一样。

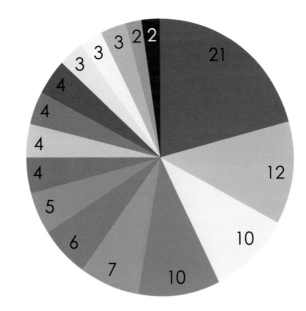

Like being wrapped in a fresh towel on the beach or donning a summer dress for café hopping or boutique strolling in the South of France. The wearer, whether male or female, will be accompanied by the response to such colourfulness.

看到这些丰富的色彩，不管是男性还是女性，或许脑海中都会出现这样的场景：披着鲜艳的沙滩巾漫步在沙滩上，或是穿着清新夏服进入法国南部咖啡馆，或是闲逛着精品店。

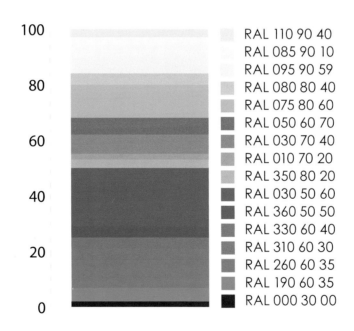

RAL 110 90 40
RAL 085 90 10
RAL 095 90 59
RAL 080 80 40
RAL 075 80 60
RAL 050 60 70
RAL 030 70 40
RAL 010 70 20
RAL 350 80 20
RAL 030 50 60
RAL 360 50 50
RAL 330 60 40
RAL 310 60 30
RAL 260 60 35
RAL 190 60 35
RAL 000 30 00

Liveliness is not introverted. Everything radiates towards the outside. Shapes release ideas, which are set in motion.

所有的笔画都呈发射状，显得活泼而不内向。这些似乎在移动的图形就像是发散的思维。

LIVELY 活泼灵动

As soon as a word can be connected with activities, red-tinged colours are in great demand. Yellow and blue variants appear only to serve as an accompaniment. Since liveliness is considered a positive feature, all dull or overly bright nuances are absent.

只要是与活动有关的主题词,就会出现许多深浅不同的红色。此处,黄色调和蓝色调充当了衬托色彩。"活泼灵动"色系具有积极向上的含义,因此图中没有暗淡或过于明亮的色彩。

/ Chapter 4

CREATIVE

天马行空

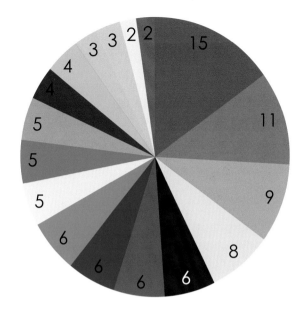

As you can see: brown is missing and beige is missing; otherwise, all the other shades are present. The large share of purple and scarlet hues hardly comes as a surprise.

图中色彩非常丰富，却显然缺失了棕色和米色，不足为奇的是紫色和鲜红色占了很大比例。

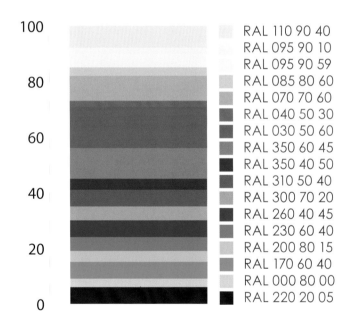

The range features a high level of colour intensity with frequent crayon-coloured shades. The creative lives off exaggeration and thus off loud signals.

图中最显著的特征就是色彩密集，而且多次出现了蜡笔色。这也说明创新与夸张、花哨的并无关联。

RAL 110 90 40
RAL 095 90 10
RAL 095 90 59
RAL 085 80 60
RAL 070 70 60
RAL 040 50 30
RAL 030 50 60
RAL 350 60 45
RAL 350 40 50
RAL 310 50 40
RAL 300 70 20
RAL 260 40 45
RAL 230 60 40
RAL 200 80 15
RAL 170 60 40
RAL 000 80 00
RAL 220 20 05

Many drawing patterns serve as symbols. When interpreting with graphic expression, the interpretation of the term knows no bounds.

许多图案都像标志，而在阐释其中的含义时，就会发现"创新"永无止境。

CREATIVE   天马行空

Creativity represents an exuberance of feelings: extremely colourful, strongly patterned, barely ordered, i.e. chaotic and rarely plain-coloured. Creativity is hardly ever characterised by reduction, but almost always by variety.

极其丰富的色彩、棱角鲜明的图案、杂乱无章的形状，这种创新体现了情感的丰富性。图中色彩混杂在一起，素色极少。创新往往不以精简为特征，而是多样性。

/ Chapter 4

IMAGINATIVE

想入非非

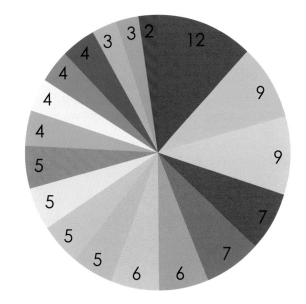

Almost twenty percent are red-tinged, only then followed by soft pastel hues. About fifteen percent consist of blue and green mid-shades.

色盘中的红色调几乎占了20％的比例，并伴有一些柔和的色彩，中性蓝和中性绿占有约5％的比例。

A bar diagram without surprises. The wealth of imagination appears to be a rather colourful mixture. Alternatively, "unimaginative" manifests itself with black and grey.

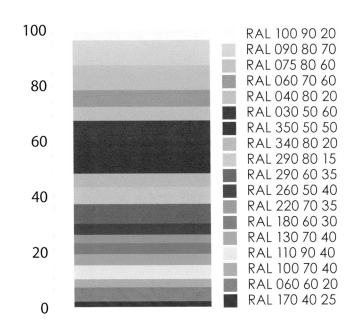

RAL 100 90 20
RAL 090 80 70
RAL 075 80 60
RAL 060 70 60
RAL 040 80 20
RAL 030 50 60
RAL 350 50 50
RAL 340 80 20
RAL 290 80 15
RAL 290 60 35
RAL 260 50 40
RAL 220 70 35
RAL 180 60 30
RAL 130 70 40
RAL 110 90 40
RAL 100 70 40
RAL 060 60 20
RAL 170 40 25

这组色彩条形图并没有出奇的效果。丰富的想象力似乎就是丰富的色彩组合。相反，"缺乏想象力"的代表色就是黑色和灰色。

Serpentine lines, spirals and a few graphic signs. They suffice for the illustration. On top of it all, row patterns and lines in various arrangements can be discovered.

  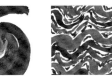

蛇形线条、漩涡以及部分图形符号充分诠释了主题。仔细观察便能发现排列顺序杂乱的横线条。

IMAGINATIVE 想入非非

The shades have many childlike features. In the eyes of many people, imagination seems to be a privilege of children. Beuys' statement "Everybody is an artist" would have an even greater claim to truth, if the quotation were: "Every child is an artist".

这些色彩有着天真烂漫的特征。在许多人的眼中，想象似乎是孩子的特权。博伊斯曾说过："每个人都是艺术家"。倘若把这句话变成"每个孩子都是艺术家"，就更贴近现实了。

/ Chapter 4

IMPRESSIVE

惹人注目

The circle is reminiscent of decorations from the military world or the carnival. It was essential to them that they outshine anything in their vicinity.

色盘让人联想到军事世界或狂欢节会中的装饰物。对于这些物品来说，比周围的一切更炫目尤为重要。

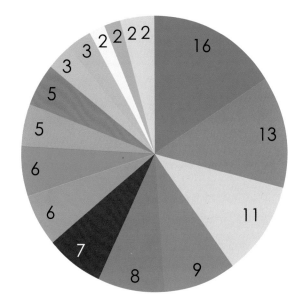

The plumage of many a tropical bird, royal parlours, fancy night bars or most expensive, patterned high heels look like this

热带鸟儿、皇家客厅、炫彩夜吧，以及昂贵的花纹高跟鞋，不正是如此吗？

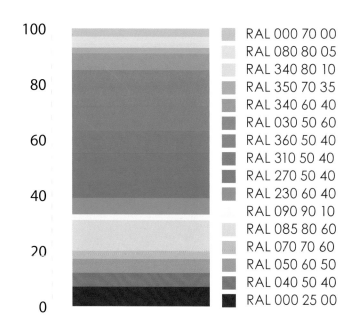

RAL 000 70 00
RAL 080 80 05
RAL 340 80 10
RAL 350 70 35
RAL 340 60 40
RAL 030 50 60
RAL 360 50 40
RAL 310 50 40
RAL 270 50 40
RAL 230 60 40
RAL 090 90 10
RAL 085 80 60
RAL 070 70 60
RAL 050 60 50
RAL 040 50 40
RAL 000 25 00

The signs of this term are relatively free of frills. Conciseness with large-volume contents is preferred.

相对来说，此主题下的图案落落大方、简洁明了，不带任何虚饰。

IMPRESSIVE   惹人注目

Colours like the uniforms of brave hussar officers at the emperor's ball at Schönbrunn Palace. An infinite number of costume dramas were condensed into such a display of colour. Shining and iridescent, from a time when the continuation of politics was transferred to the operetta stage.

美泉宫中国王舞会上勇敢的轻骑兵所穿的制服，也有着此般艳丽的色彩。此处，无数部古装戏剧被浓缩成一场色彩盛宴，又像政治舞台被转换成轻歌剧舞台的那一刻，光芒四射、熠熠生辉。

LIVELY
活泼灵动

CREATIVE
天马行空

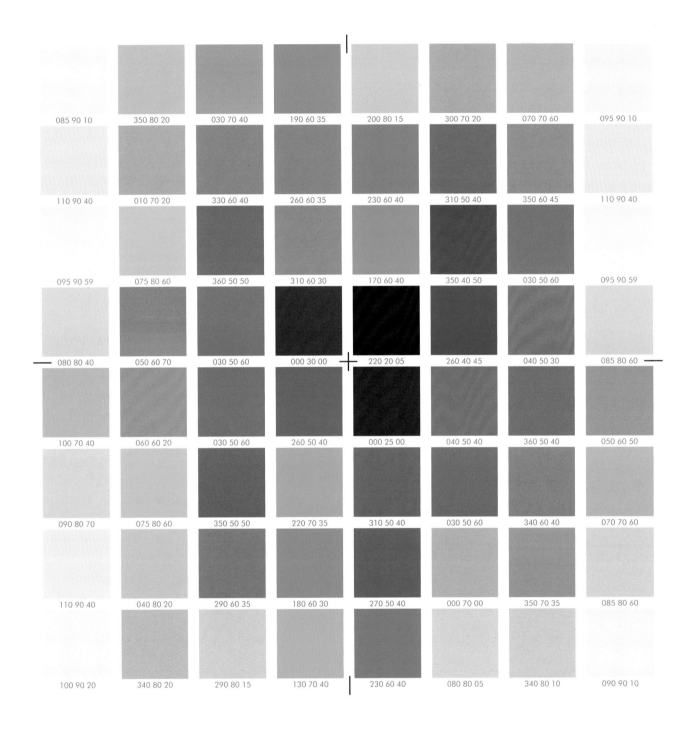

IMAGINATIVE
想入非非

IMPRESSIVE
惹人注目

LIVELY
CREATIVE
IMAGINATIVE
IMPRESSIVE

活泼灵动
天马行空
想入非非
惹人注目

Four themes and a similar dynamism. The stimulating power of the terms is reinforced by bright and dark or cold and warm accents. Every adjective accordingly has a larger share of sunny shades than watery blue tones. The colour scheme contains a large number of playful and musical elements. It can be imagined as a model for a melodic sketch as well as for a complete interior design ensemble.

Beautiful rose, lilac and violet shades are present, as are shades of icing and chocolates. The large number of sunny yellow and golden hues is surprising. Summery flower and blossom tones also abound, and they leave behind the strongest and most inspiring impression.

四个主题下的色彩有着同样的活力，而深、浅色调和冷、暖色调的并用增强了其力量感。在每个形容词之下，阳光色总是多过水蓝色。这组色彩搭配方案中包含了许多趣味元素和音乐元素，可被想象成旋律优美的曲谱，或是完美的室内设计。

迷人的玫瑰色、丁香花色和紫罗兰色，以及深浅不一的糖衣色和巧克力色都呈现在眼前；多种多样的日光黄和金色更是令人惊奇万分；夏花盛开，色彩娇艳，令人印象深刻，引发无限想象。

IMAGINATIVE 想入非非

Patterns arouse people's imagination. They captivate the eyes. Furthermore, they are guiding models with a symbolic character, which, if they are derived from real examples, convey a choreography of order and systematisation. Patterns can stimulate and soothe; they can be light and buoyant or static and graphical, network-like or linear.

Once more, patterns are in great demand. A promi-sing future is predicted for the decorative, which was long frowned upon, because we have abandoned large lifeless surfaces. A better form of entertainment for walls, floor and to some extent also for ceilings are lively amorphous shapes or slightly more expressive structures, and the even more distinctive scattered and grid patterns – more than expressive flower decors with large repeats or mixed patterns, the form of which derives from flora, fauna and landscape references.

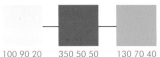
100 90 20   350 50 50   130 70 40

这些图案令人浮想联翩，目眩神迷。它们更是个性鲜明的指导模板，如果是作为实例的反映，则更具有类似舞蹈动作设计的顺序性和组织感。图案可以刺激人心，也可以抚平人心；其色彩可以浅淡积极，也可以沉着生动；可以呈网状，也可以呈线状。

此处又需各式花样。长久以来都被人批判的装饰，在去掉毫无生机的表面装饰之后，终于又显得生机勃勃。墙壁、地面甚至天花，有了自由生动的形状、引人注目的结构、独特的散乱或网格图案后，会更赏心悦目，而不断重复、交织的动植物和风景图案则难以达到这种效果。

340 80 20   350 50 50

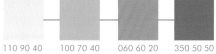

/ Chapter 4

INSPIRING

心驰神往

The pie chart, illustrates how buoyantly the colours can stage a dynamic effect. About half of the shades could be characterised as either warm or cold.

色盘充分阐释了鲜艳积极的色彩是如何制造动态效果的，其中约有一半的色彩可归为暖色或冷色。

A very pleasant looking range presents itself here: it is light, feminine and pure-toned, fashionable rather than factual, floral rather than faun-like and active rather than passive.

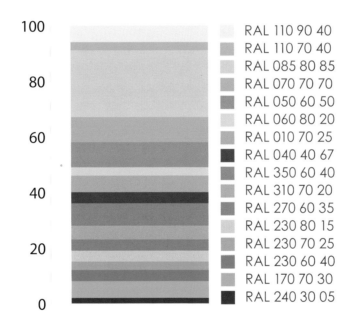

RAL 110 90 40
RAL 110 70 40
RAL 085 80 85
RAL 070 70 70
RAL 050 60 50
RAL 060 80 20
RAL 010 70 25
RAL 040 40 67
RAL 350 60 40
RAL 310 70 20
RAL 270 60 35
RAL 230 80 15
RAL 230 70 25
RAL 230 60 40
RAL 170 70 30
RAL 240 30 05

这组美丽的色彩有着浅淡、娇柔、纯然、时尚的感觉，而没有真实感；有着鲜花般的绚烂，而没有农牧之神的神秘；积极主动而非消极被动。

The drawing style gets by with pure abstractions: stripes, circles, dots, curves, and here and there a few hints of leaves and fruits.

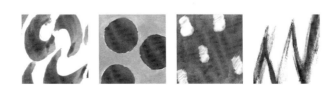

条纹、圆圈、圆点以及曲线，有时又像树叶和果实，体现出纯抽象的绘画风格。

INSPIRING　　　　心驰神往

The shades look cheerful, playful and contagiously funny. There is virtually no sign of a gloomy mood. The few dark lines serve only to help to accentuate a painted square.

这些色彩显得欢快、俏皮且趣味十足，而不带一丝忧郁。几条黑色线条仅仅充当了凸出方形画板的角色。

MOTIVATING

激发斗志

The cold and warm shades used here generate a range of complementary contrasts: shades like red and green, yellow and blue, which are positioned opposite each other in the colour circle.

冷、暖色调对比鲜明，又有互补的作用，如色盘中红、绿、黄、蓝四种色彩都处于对立的位置。

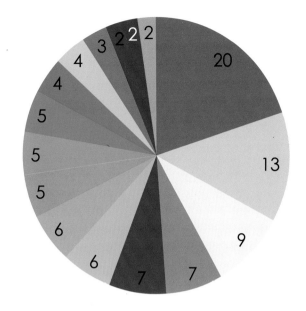

This colour sequence is strongly emphasised. The pure shades affect our consciousness in a similar way to loud sound sequences or sweet-and-sour dishes with odour and flavour enhancers.

该纯色系色彩十分强烈，就像嘹亮的声音，或添加了增味料的酸甜菜肴，刺激着我们的感官。

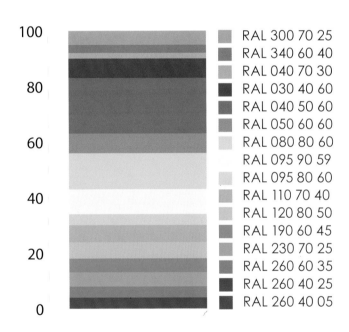

- RAL 300 70 25
- RAL 340 60 40
- RAL 040 70 30
- RAL 030 40 60
- RAL 040 50 60
- RAL 050 60 60
- RAL 080 80 60
- RAL 095 90 59
- RAL 095 80 60
- RAL 110 70 40
- RAL 120 80 50
- RAL 190 60 45
- RAL 230 70 25
- RAL 260 60 35
- RAL 260 40 25
- RAL 260 40 05

The shapes reveal a simple sign language. The colour scheme is the more important means of communication.

图案的形状就像是简单的符号语言，而用色才是传递信息的重要方式。

MOTIVATING 激发斗志

With great verve, the primary colours convey energised bursts of motivation. Accordingly, red is the catalyst for propulsion. Plain colours are frequently used. The surfaces are mostly free from mixed colours. Sometimes we can also recognise hazy, slightly blurred hues.

这些原色显得热情洋溢，就像是动力的有力爆发，而红色就是激发动力的催化剂。当然，素淡的色彩也出现次数较多。总体来看，没有过多的混合色彩，当中浅淡模糊的彩色也依稀可见。

ENTERTAINING

娱乐至上

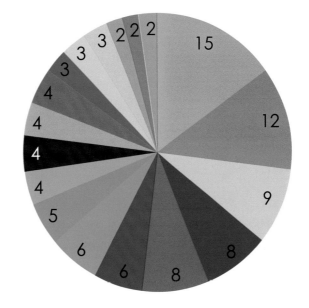

Bold crayon colours encircle the rainbow three times! These are the shades in which one would like to bathe and relax. Here, grand opera, bibulous operetta and kitschy musical are combined in a single performance.

大胆的蜡笔色彩就像三道彩虹，出现在色盘上。这些就是人们想沐浴、休憩于其中的色彩，似乎融合了大歌剧的壮观、轻歌剧的欢快以及通俗音乐剧的流畅。

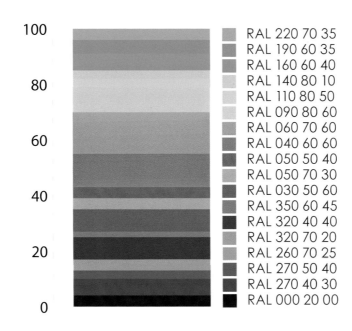

Colour bars as a dynamic visual treat. Colours for all those with a perpetually child-like spirit, an intellect free from arrogance as well as plenty of sympathy for exaggeration and contempt for timid simplicity.

这组彩色条演绎了一场动态的视觉盛宴。这些色彩专属于永怀童心、充满智慧而不傲慢的人，以及怜悯忧天杞人、轻视懦弱小人的人。

Every little dot that plays with the other dots is entertaining. It is equally great fun to put one colour next to another and see how they enjoy themselves.

圆点与圆点互相作用，充满趣味。而把两种色彩并列起来，看它们如何争艳，也同样有趣。

ENTERTAINING   娱乐至上

This is pure funfair, circus, entertainment, and a superb hotel experience! Wild colours and shapes are part of the scene, and so are sport, play and dance. White is stationary, grey is taboo, and not even the night is black because neon plays along.

这恰似在玩转游乐场、欣赏马戏团表演，或者体验顶级酒店，令人激动无比。自由的色彩、自由的形状，就像以上场景中的一部分，又像是运动、游戏和舞蹈中的一部分。白色代表静止，灰色代表禁忌，但夜晚在霓虹灯的照耀下，并非是黑暗的。

TEMPTING

鲜艳夺目

Not only red and purple but also pink, rosé, violet, lilac, peach, lavender, violet, and cream hues turn our heads. However, they are garnished with a few hints of black and pure aqua.

色盘中不只有红色和紫色，也有粉红色、玫瑰色、紫罗兰色、丁香花色、桃红色、薰衣草色以及奶油色，还点缀着些许黑色和浅绿色。

These sky colours are more exciting than any sunset no matter how splendid it is. If nature is kind to us every once in a while, we shall see this spectacle of rainbow colours.

落日的色彩不管有多辉煌艳丽，都没有天空的色彩那么动人。偶尔大自然也会展现她的善良，让我们看看彩虹那美丽的衣裳。

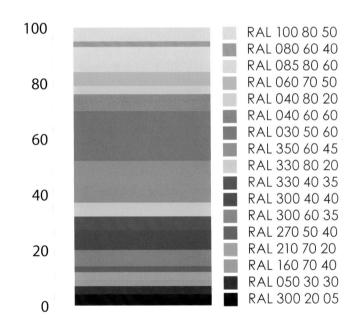

Frequently, it is the round or concentric signs, as well as the serpentine lines, which attract people. Zigzag patterns or diagonal stripes are not among the favourite shapes.

通常，圆圈（或同心图形）和蛇形线条会受人喜爱，而锯齿图案和对角线则不然。

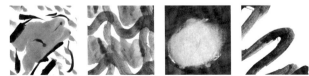

TEMPTING 鲜艳夺目

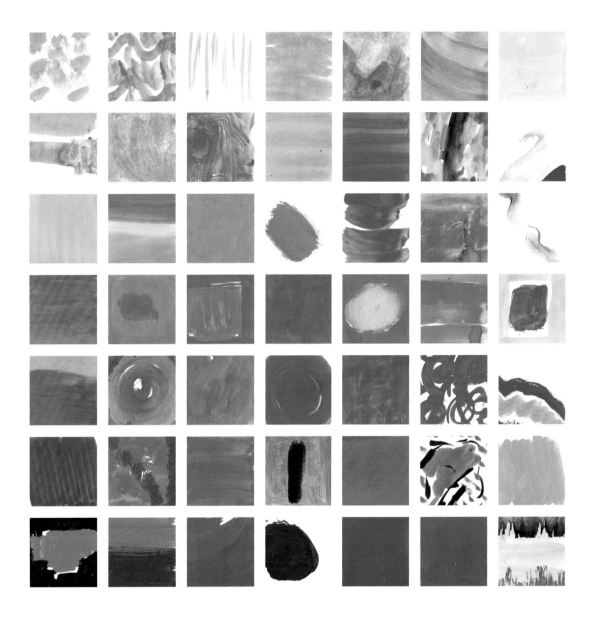

Temptation strategies seldom succeed with shades of brown, grey and beige. The mystery of sweet allure, seductive luxury and pleasurable thrill is far removed from any mere invitation. Such colours do not conceal but openly reveal the energetic seduction.

棕色、灰色和米白色往往难以吸引人的眼球，而神秘诱人的甜蜜色、魅力十足的华丽色以及愉悦人心的热烈色，它们并不收敛自身的诱惑力，而是将其毫不隐藏、大方地展现出来。

INSPIRING
心驰神往

MOTIVATING
激发斗志

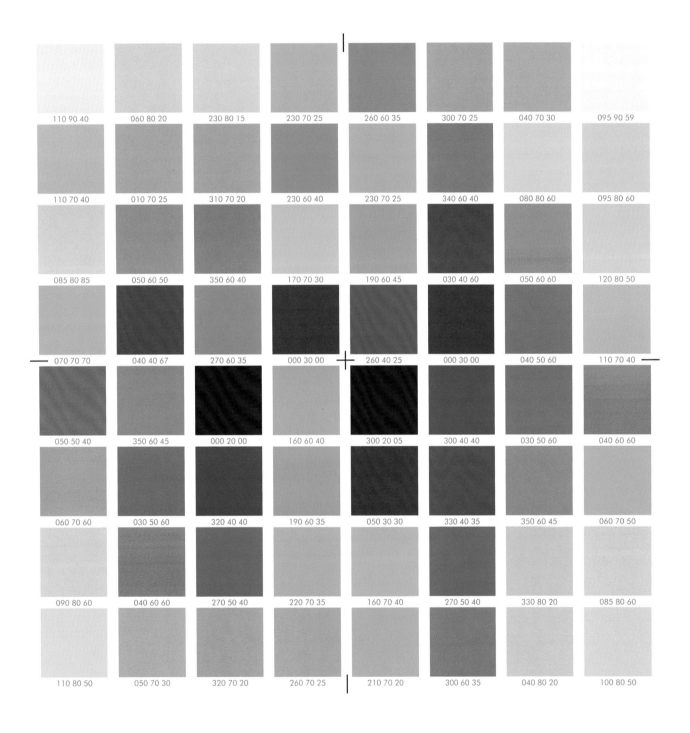

ENTERTAINING
娱乐至上

TEMPTING
鲜艳夺目

INSPIRING
MOTIVATING
ENTERTAINING
TEMPTING

心驰神往
激发斗志
娱乐至上
鲜艳夺目

The entire field is a feast for the eyes, based on overemphasis. Here, nothing has been created with calm solidity or kindest regards. Theatrical colour demeanour is married with tempting offers, and loud motivation leads to inspiring colour canons. If we take another look back at the original interpretations, we become aware that the original drawings are making a statement even more emphatically than the subdued summary is able to.

We know that we can live perfectly well with these colours. From time to time we may need a visual respite, which may subdue the sorbet hues to the pale limestone shades shown here. Some of the brighter shades of blue should also be used as recreation colours in a brightened, additionally greyed and thus soothing way.

这里没有沉静死板、柔和温暖的色彩，整片色域色彩鲜艳，令人大饱眼福。强烈夸张的色彩属于"鲜艳夺目"区；活力四射的色彩属于"心驰神往"区。如果把最初的阐释带入其中，我们会发现，原来的图画会比此处的色彩总盘更具表现力。

生活中充满这些色彩会十分完美，但有时我们的眼睛也需要休息，因此此处将鲜艳的冰糕色替换成了暗淡的石灰石色。而且，亮蓝色变淡、变灰之后，也能令人身心放松。

INSPIRING   心驰神往

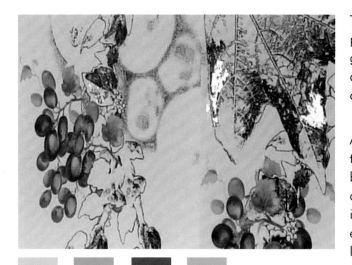

The generation-spanning atmosphere calls for lively and pleasant conversations. Intimate tables for two or for small groups encourage intimate gatherings. The creative impression and the colours are indicative of a relatively light, exquisite cuisine.

An ingeniously staged ambience establishes contacts between the guests because they remain visible and are not hidden behind decorative objects or partitions. Unobstructed visual axes in the interior are popular since we are able to participate in the great pleasure of people getting to know each other, even though the encounters may often be only fleeting. The longer the guests stay, the higher is the appreciation of a clear all-around view. The small table lamps provide the guests with intimate privacy.

060 80 20   010 70 25   040 40 67   070 70 70

040 40 67   060 80 20   050

　　多代共用的空间应能让人积极地交谈起来，并让人从中获得快乐；适合两人或者多人的桌子应能满足亲密聚会的要求；创意和色彩则应让人想到清淡精致的美食。

　　精心营造的气氛有利于顾客之间建立联系，因为顾客之间都是互相可见的，而非隐藏在装饰物品或隔屏之后。在视线毫无阻隔的室内空间中，我们能与人群同享快乐，互相了解，即使只是匆匆过客。

170 70 30   110 70 40   070 70 70   060 80 20

　　这种格局的空间越来越受欢迎，而且顾客逗留得越久，就越欣赏这种开阔的视野。小台灯更能拉近顾客与顾客之间的距离。

350 60 40   060 80 20

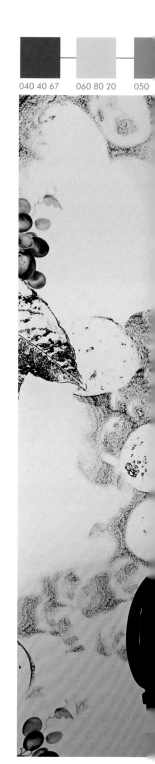

060 80 20   085 80 85   040 40 67

060 80 20   085 80 85   070 70 70

INSPIRING

/ Chapter 4

MULTI-CULTURAL

多元文化

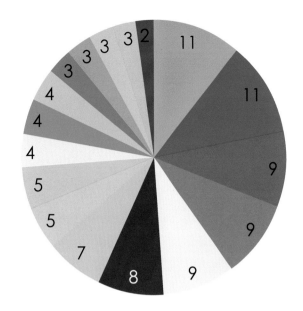

The multifacet aspects of the tonal interpretations indicate an immense volume of associations. Multi-cultural also means an excess of colour diversity.

多层次、多角度的色调暗指巨大的关系网。"多元文化"也指色彩的多样性。

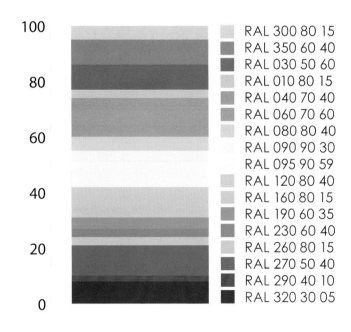

RAL 300 80 15
RAL 350 60 40
RAL 030 50 60
RAL 010 80 15
RAL 040 70 40
RAL 060 70 60
RAL 080 80 40
RAL 090 90 30
RAL 095 90 59
RAL 120 80 40
RAL 160 80 15
RAL 190 60 35
RAL 230 60 40
RAL 260 80 15
RAL 270 50 40
RAL 290 40 10
RAL 320 30 05

The colour stripes display the overview in a more orderly fashion than a pie chart could ever do. Mystic violet nuances and all shades of brown are missing.

相对扇形图而言，此条形图更能清晰地呈现整体的效果。神秘的紫罗兰色调和棕色调并未出现在图中。

The drawing style features horizontal or vertical layers: the multi-cultural as an organisational structure. Circles and little dots show the playful and dynamic reference points for empathy.

图案把"多元文化"比作组织结构，因此横向和纵向的层次感尤为显著。圆圈和小圆点则体现了"移情"的趣味性和动态感。

MULTI-CULTURAL 多元文化

Did not we know it all along? "Multi-cultural" means colourful. Not even ten percent of the study participants contented themselves with a one-colour presentation. Almost without exception, the shades are mixed in a stimulating way. The colour commitment is unambiguous, positive and energised.

"多元文化"意指色彩缤纷，难道我们向来都不知道吗？不到10%的研究参与者满足于使用一种色彩，然而，几乎所有人都会把鲜艳的色彩交织起来。这些色彩显然清晰、积极、饱含活力。

/ Chapter 4

INTERNATIONAL

国际风范

A hint of grey or black is also included as a secondary colour shade. The accompanying colours range right through to luminous, pastel-toned hues.

从明亮色调直接过渡到浅淡色调，并有少量的灰色和黑色作为次要色调出现在色盘中。

The bar chart contains buoyant shades. The result is a playful colour scheme that appeals to the young. The "international" aspect indicates a stable, rather than a short-term politically coloured illustration of the term.

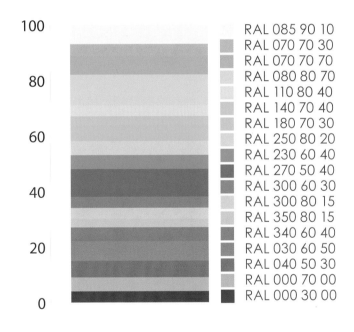

RAL 085 90 10
RAL 070 70 30
RAL 070 70 70
RAL 080 80 70
RAL 110 80 40
RAL 140 70 40
RAL 180 70 30
RAL 250 80 20
RAL 230 60 40
RAL 270 50 40
RAL 300 60 30
RAL 300 80 15
RAL 350 80 15
RAL 340 60 40
RAL 030 60 50
RAL 040 50 30
RAL 000 70 00
RAL 000 30 00

此色条图是由许多鲜艳的色彩组成的，显得十分俏皮，是年轻人的最爱。"国际风范"指的是稳定而非短暂的具有政治意味的色彩。

Mosaic and stripy shapes constitute the main repertoire of patterns. As we can see, large amounts of shape and colour types are necessary.

四幅图案中马赛克和条状图形最为突出，当然，多种多样的图形、深浅不同的色彩，也都不可或缺。

INTERNATIONAL　　国际风范

Things are colourful and peaceful. The term has positive quality features. The colours are highly motivating but partly also placating: yellow and orange as tones full of atmosphere and violet, blue and green as soothing colours.

万物缤纷多彩、安详平静，这就是图中的景象。"国际风范"有着积极的意义，因此相应的色彩也大都鲜艳浓烈，只有少部分柔和色彩，如充满活力的黄色和橘色，以及柔和的紫色、蓝色和绿色。

CLASSIC

经典传承

The colour emphasis points towards elegance, function and balance. The technical grey-black and blue shades outweigh the warm, earthy ones.

庄严稳重的灰黑色和蓝色，远远多过暖色和素色，这组色彩旨在突显优雅感、功能性和平衡感。

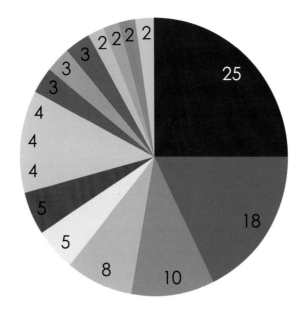

All in all, the study participants prefer a colour range based on reason and solidity: it is interesting that the "classic" colour profile represented here only displays partial relationships with its historic origins.

在该主题下，研究参与者更倾向于理性和稳健的色彩。有趣的是，此处只展示了经典色系与历史渊源的部分联系。

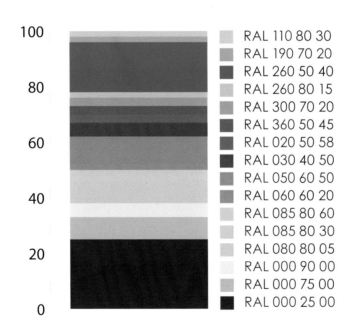

RAL 110 80 30
RAL 190 70 20
RAL 260 50 40
RAL 260 80 15
RAL 300 70 20
RAL 360 50 45
RAL 020 50 58
RAL 030 40 50
RAL 050 60 50
RAL 060 60 20
RAL 085 80 60
RAL 085 80 30
RAL 080 80 05
RAL 000 90 00
RAL 000 75 00
RAL 000 25 00

The shapes are graphical, straight and predominantly symmetric. There is no room for dynamic effects.

规矩的图形、笔直的线条以及对称的图案，没有丝毫动感。

CLASSIC   经典传承

The classic is situated somewhere between symmetry and harmony, between regularity and balance. Similarly, it is well-shaped or well-proportioned. This is not only reflected in the interpretation with shapes but equally in the colour scheme.

经典是处于对称与和谐、规整与平衡之间的，而且具有形状匀称、比例协调等特征。这不仅在图形中有所体现，在配色方案中也十分明显。

FOLKLORISTIC

地域风俗

In their mind's eye, many people probably see the same picture as that represented here. Expectation and reality fit together perfectly.

大多数人们所看到的色彩，或许与他们对"地域风俗"的联想一致。这正是现实与期望的完美结合。

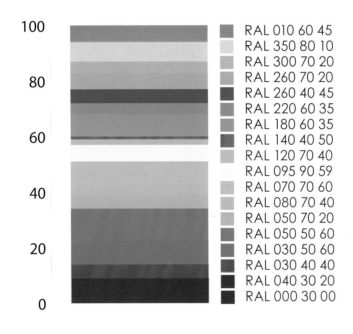

This sequence is quite substantial. When looking at it carefully, about sixty percent of the hues are, just like the contents of the drawing, coloured playfully.

RAL 010 60 45
RAL 350 80 10
RAL 300 70 20
RAL 260 70 20
RAL 260 40 45
RAL 220 60 35
RAL 180 60 35
RAL 140 40 50
RAL 120 70 40
RAL 095 90 59
RAL 070 70 60
RAL 080 70 40
RAL 050 70 20
RAL 050 50 60
RAL 030 50 60
RAL 030 40 40
RAL 040 30 20
RAL 000 30 00

这组色彩整体看来相当饱满，但细看就会发现，浓艳的色彩约占了60%，就像整幅图片呈现出来的那样，多彩而有趣。

The folkloristic features are not exactly characterised by restraint and modesty. In an onomatopoeic and powerful fashion, they make a beautiful racket.

"地域风俗"的特征并不一定是保守和朴素。线条如声音般有力地结合起来，组成美丽的旋律。

FOLKLORISTIC　　　　地域风俗

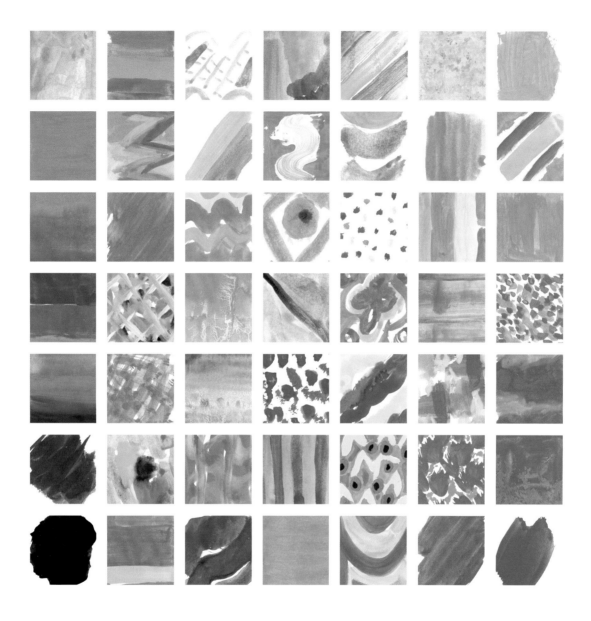

The gay colours of traditional dress are in the majority. Earth and forest shades take a back seat. The more rural the homeland idyll, the more brown-dark red- and olive green-bound is the colouring.

全图以传统礼服般的鲜艳色彩为主，其次就是大地和森林的色彩，再者就是偏向于乡村和田园风格的暗棕红色和橄榄绿。

MULTI-CULTURAL
多元文化

INTERNATIONAL
国际风范

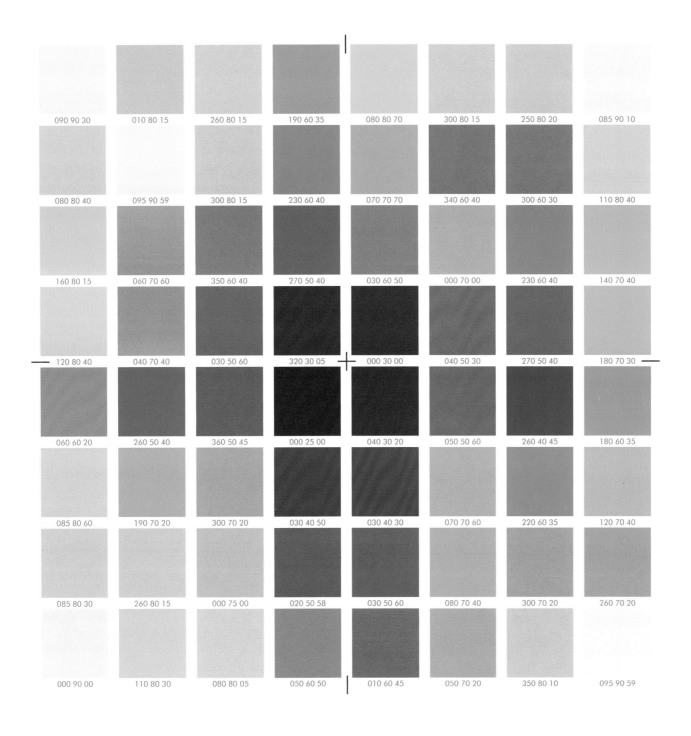

CLASSIC
经典传承

FOLKLORISTIC
地域风俗

MULTI-CULTURAL
INTERNATIONAL
CLASSIC
FOLKLORISTIC

多元文化
国际风范
经典传承
地域风俗

All four terms make reference to potentially different design temperaments; the objective, however, always aims at design structures that are independent of each other ethnically and geographically as well as culturally and temporally.

The colourings are not particularly diverging, as might have been assumed. Their profiles reveal a common conclusiveness of diverse unity and richness. Only the folkloristic colour characteristics show an admixture of warmer shades of about five percent. We should also recognise that hybrid forms of classic-folkloristic or multi-cultural-classic ambient settings can be appealing.

This range of variation in particular is a welcome example of how what was originally heterogeneous can be harmonised to create a beautiful bel canto of colour.

　　四种配色主题下的设计效果会完全不同，而各个主题的宗旨在于设计在种族、地理、文化和世俗等方面都不相同的结构。

　　或许这些色彩并不是太分散，总体显得丰富多样而又统一。只有"地域风俗"主题中混合了多种暖色，占了约有5%的比例。还应注意的是，"经典传承"分别与"地域风俗"和"多元文化"搭配，效果也会十分迷人。

　　这组色彩具有多样性，恰到好处地阐释了各种艳丽的色彩是如何达到相互协调，最终组成美丽的色盘的。

CLASSIC 经典传承

The issue of establishing the truth as to what "classic" means, is taught us by the numerous examples and models which, originating from Greco-Roman antiquity, keep on renewing themselves so that they continue to demonstrate their style-forming effect to this day. On that subject, literature states that the classic was understood in imitative times as a normative paradigm: so, among others, the golden age of the Hohenstaufen dynasty: Staufer classic, around 1200, then the Classicism of the Age of Goethe from 1770 to about 1830. Classic has, of course, originally to do with "class". It is essential that balance and symmetry, rather than asymmetry, retain the upper hand, never agitation or chaos.

From perceptual psychology we know that the mirror-imaged face is without exception considered more harmonious than the real one. Taking the middle in a group of three or five is more valuable than a position on the edge. A row of identical parts at identical intervals is always more successful as a design type than a single product – as illustrated on the reception counter with the three stately lamps.

这些源自古希腊和古罗马的大量案例和模板，向我们阐释了如何确认真相以及如何理解"经典传承"，它们不断更新，将这些风格的影响力延续至今。就这个问题有文献指出，在模仿时代中，经典被理解为一种标准范式，如1200年左右，霍亨斯陶芬王朝鼎盛时期的斯陶芬经典，还有1770至1830年左右的歌德时代经典。显然，经典最初与阶级有关。因此，相对动荡、杂乱、不对称而言，平衡、匀称与之更贴切。

从感知心理学的角度讲，镜中映照的一个人的样貌，无一例外地都比他的真实长相更加耐看。再比如三五一组，中间的位置通常比靠边的位置更具重要性。有序排列的产品组合，也永远比单一的产品更加畅销，这正好解释了接待台为何并列设有三盏明灯。

000 90 00   000 75 00   020 50 58

030 40 50   000 75 00   000 25 00

For this reason, too, the classical design concept is so convincing because it has a high level of aesthetic effectiveness. It corresponds to our pronounced sense of harmony and calm. Slants, zigzag and multiple intersections of lines and shapes should be avoided. The design of suites and other rooms illustrates how agreeable the formal design but also the calm and yet not unaccented handling of the applied colours can be. Shades as well as colours should always be repeated: the red wall colour as well as the red cushions and the colour of the three lampshades is found again several times around the counter.

由于经典设计有更高的美学效果，因此这种设计理念更有说服力，它符合我们对和谐与平静的感知。而应当避免的是斜线、"之"形线以及繁复的交叉线。套房和其他房间的设计表明，条理清晰、色彩朴素沉静的空间是何等惬意。形状和色彩则应重叠使用，如墙壁和沙发重复使用了红色，三盏灯的灯罩与柜台也采用了同样的色彩。

000 75 00
085 80 30
000 25 00
000 90 00
300 70 20

000 25 00
000 90 00
000 75 00

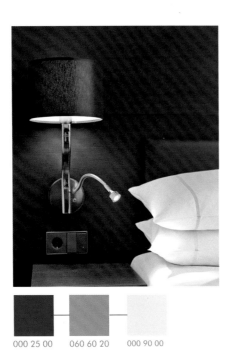

000 25 00	060 60 20	000 90 00

000 25 00

085 80 60

000 75 00

000 90 00

085 80 60	030 40 50	000 25 00

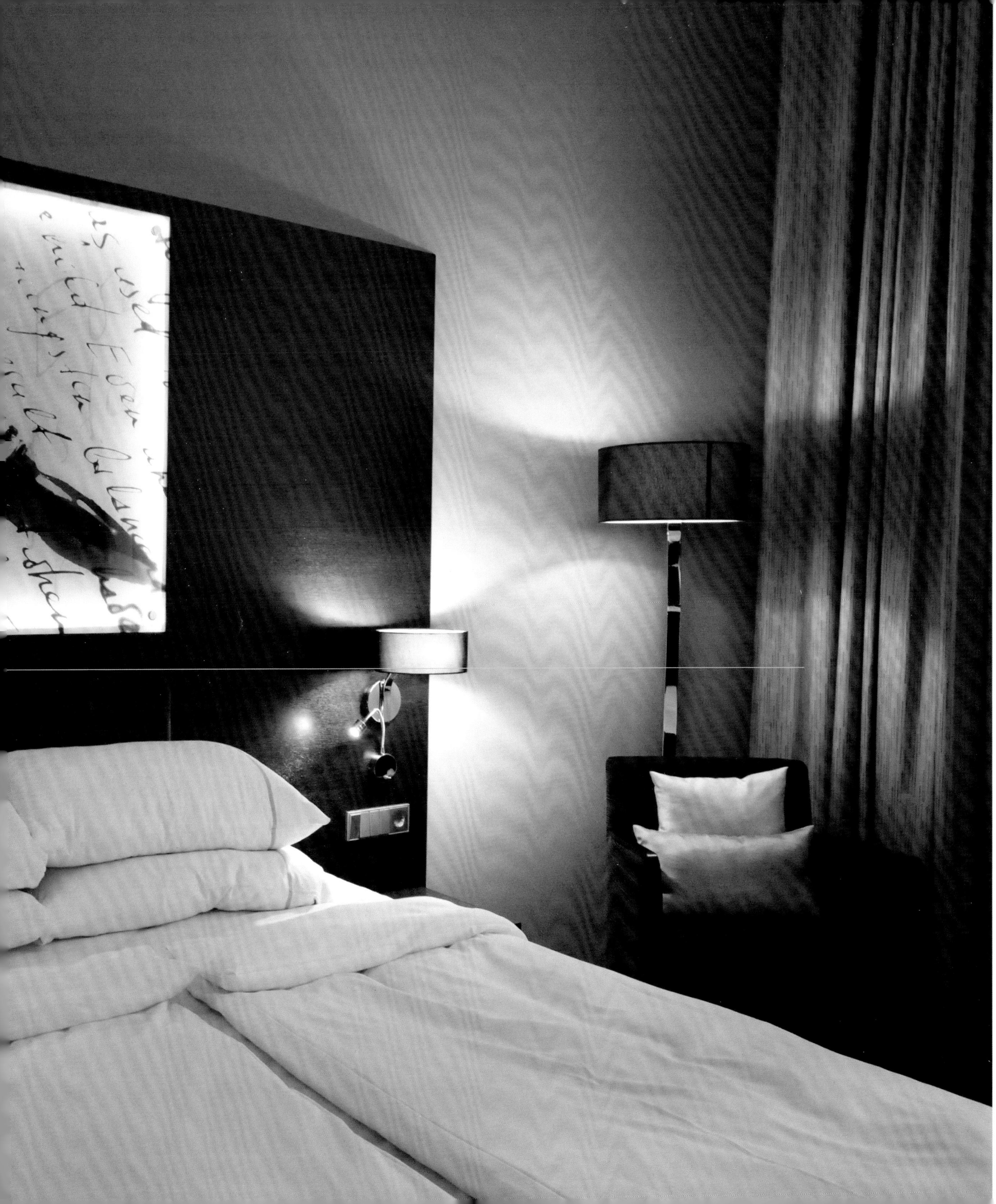

FAIRYTALE-LIKE

恍如童话

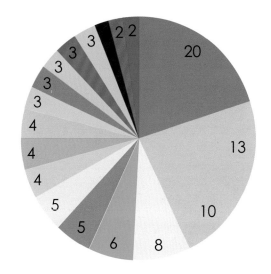

The pie chart tastes nice. It is wonderfully sweet and properly garnished with fruit flavour. The ideal colour world needs large amounts of pink, lilac, peach, sea-green, and ice cream hues.

色盘里的色彩如此甜美，甜似水果。理想的用色方案包括大量的粉色、丁香紫、桃红、海绿色和冰激凌色调。

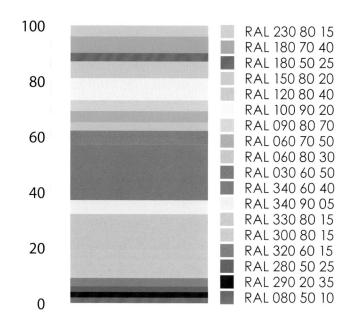

RAL 230 80 15
RAL 180 70 40
RAL 180 50 25
RAL 150 80 20
RAL 120 80 40
RAL 100 90 20
RAL 090 80 70
RAL 060 70 50
RAL 060 80 30
RAL 030 60 50
RAL 340 60 40
RAL 340 90 05
RAL 330 80 15
RAL 300 80 15
RAL 320 60 15
RAL 280 50 25
RAL 290 20 35
RAL 080 50 10

The bar chart does not clear the colour impression of the familiar harbingers of kitsch. When the Queen goes out towin over the crowds in pastel-motley colours, the resemblance is remarkable.

条形图呈现的色彩并非常见的流行色，走的是像女王出宫迎接民众般的高贵又亲民的路线，类似于淡彩甜甜圈的色彩。

Stars and Stripes, wave lines, blossoms, the sun, bouquets, leaves, and iridescent cloudscapes are indications of the fairytale.

图案丰富多样，比如星星、条纹、波浪、花蕊、太阳、花束、叶子、七彩祥云等形状，充满童话意味。

FAIRYTALE-LIKE　　　　　　　　恍如童话

The fairytale is occupied by fairies, good spirits and Barbie dolls. We had better keep the evil ones outside and chase them away with large amounts of pink. The nice and the capricious ones and the princesses are always more important than horrid witches and black-green mixers of poison.

　　童话故事当然得有精灵、天使和芭比娃娃。我们要把粉色当作武器，将邪恶力量排除在外。比起可怕的女巫和用黑色与绿色调制出的毒药，美丽而多愁善感的公主才是童话故事里永恒的主角。

/ Chapter 4

COSY

亲密无间

Many infantile memories are, of course, embedded in these shades. They give comfort, security and sweet dreams. The cosy corner can, in any case, be enjoyed more safely than too much cake, chocolates and liqueur.

这些色彩让我们想到记忆中婴儿时的稚气。它们意味着舒适、安全感和甜甜的梦乡。一个舒适温暖的小角落带给我们的安全感和享受，是再可口的蛋糕、巧克力和甜露酒也给不了的。

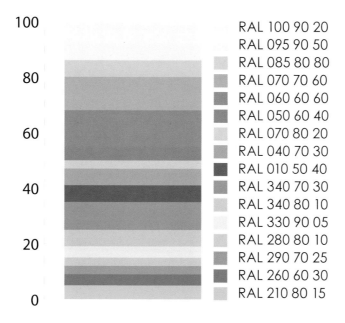

The colour sequence also signalis how amiable and desirable the cosy looks: the memory of coloured candy floss, silk cushions sweets, candy canes, and children's birthdays catches up with us.

RAL 100 90 20
RAL 095 90 50
RAL 085 80 80
RAL 070 70 60
RAL 060 60 60
RAL 050 60 40
RAL 070 80 20
RAL 040 70 30
RAL 010 50 40
RAL 340 70 30
RAL 340 80 10
RAL 330 90 05
RAL 280 80 10
RAL 290 70 25
RAL 260 60 30
RAL 210 80 15

从色序中不难领会"亲密无间"风格带给人的亲切与贴心感：像彩色棉花糖、甜心丝绸软垫、棒棒糖，又如孩子们的生日派对。

The drawings mostly consist of tone on tone areas; only rarely do they include stripes or roundish structures.

所绘图案大多有着同色异调的特点。仅有极少量的条纹或圆形结构。

COSY 亲密无间

That's how one imagines cosy armchairs with a cuddly cushion. The longing for the rosy-red, orange-yellow-sky blue sense of wellbeing becomes particularly great, when we miss these shades. Every house, every suite or every hotel room actually needs a cosy corner.

若要选择舒适的扶手椅加上可爱的靠垫，可从中找到灵感。玫瑰红、橙黄与天空蓝，都是极佳的配色方案。每幢房子或酒店套房，都需要这样"亲密无间"的舒适一隅。

/ Chapter 4

SNUG

舒适温暖

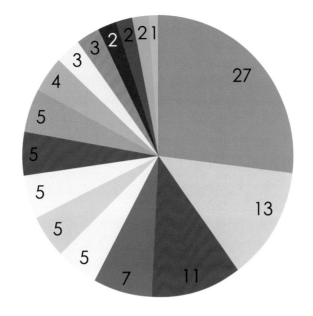

Some cosy bar parlours or guestrooms radiate this feelgood atmosphere. They are especially pleasant in winter with a crackling fire in the hearth.

从舒适的酒吧休闲厅或酒店客房中,都能找到这种给人快乐感觉的氛围。如有冬日炉膛里噼噼啪啪的柴火带来的温暖,便会更加惬意。

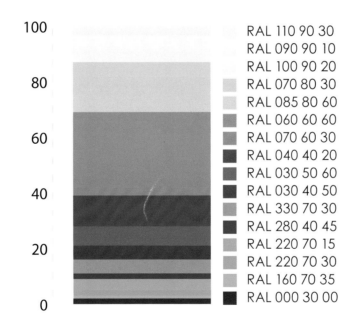

RAL 110 90 30
RAL 090 90 10
RAL 100 90 20
RAL 070 80 30
RAL 085 80 60
RAL 060 60 60
RAL 070 60 30
RAL 040 40 20
RAL 030 50 60
RAL 030 40 50
RAL 330 70 30
RAL 280 40 45
RAL 220 70 15
RAL 220 70 30
RAL 160 70 35
RAL 000 30 00

The colour range very clearly indicates that the shades originate from nature. They look woody and partly earthy. A few brighter floral spots complement the overall picture.

色彩的灵感来自大自然,呈现了木头和土地的色调,也有少许亮色点缀其间。

This term does not need large patterned examples! It is especially snug when the round form contents come to the fore.

不需要太多的图案示范,温润的圆形足以诠释"舒适温暖"。

SNUG 舒适温暖

The study participants fond of painting have developed a warming, partly lulling colour sequence. These turned out to be true feelgood colours, which generate a warming atmosphere. Most colour fields are plain-coloured.

对于这一风格,研究参与者偏爱温暖而使人平静的色彩。它们都带给人良好的视觉感受,营造出温馨的氛围。图中的色彩绝大部分为素色。

NOSTALGIC

More grey than gold, more matt than glamorous. However, the retrospective mood does without shrill shades that admonish or remind.

温婉怀旧

这些色调仿佛被蒙上了一层阴影。温婉的还旧基调，以有着伤感气息的绿锈色为主。

This sequence of tones looks fairly dusty. Its warm-toned basic colouration suggests that what we understand by "nostalgic" seems to be covered up by the plain pathos of the patina.

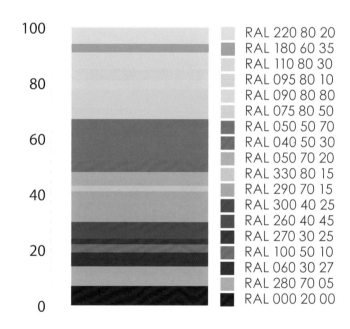

RAL 220 80 20
RAL 180 60 35
RAL 110 80 30
RAL 095 80 10
RAL 090 80 80
RAL 075 80 50
RAL 050 50 70
RAL 040 50 30
RAL 050 70 20
RAL 330 80 15
RAL 290 70 15
RAL 300 40 25
RAL 260 40 45
RAL 270 30 25
RAL 100 50 10
RAL 060 30 27
RAL 280 70 05
RAL 000 20 00

灰色多过金色，亚光色彩多过光彩夺目的色彩。还旧的心境不能忍受尖锐刺眼的亮色。

The few signs are lumpy or graphically shaped. The majority of participants practised self-denial in their graphical interpretation.

简洁的图案呈现凹凸起伏的形状。大多数研究参与者在图释中进行了自我否定的实践。

NOSTALGIC　　　　　　　　温婉怀旧

The view of the past is hazy, whether disguised nostalgically or interpreted scientific-factually. A more or less dense veil of mist covers up the past – no matter whether we look at it with quiet melancholy or uneasy feelings.

无论是出于怀旧情结还是基于科学事实，回首过去，记忆总是朦胧的。如同或薄或厚的一层面纱，蒙住过往，我们只能透过面纱看到它，安静无语，却思绪万千。

FAIRYTALE-LIKE
恍如童话

COSY
亲密无间

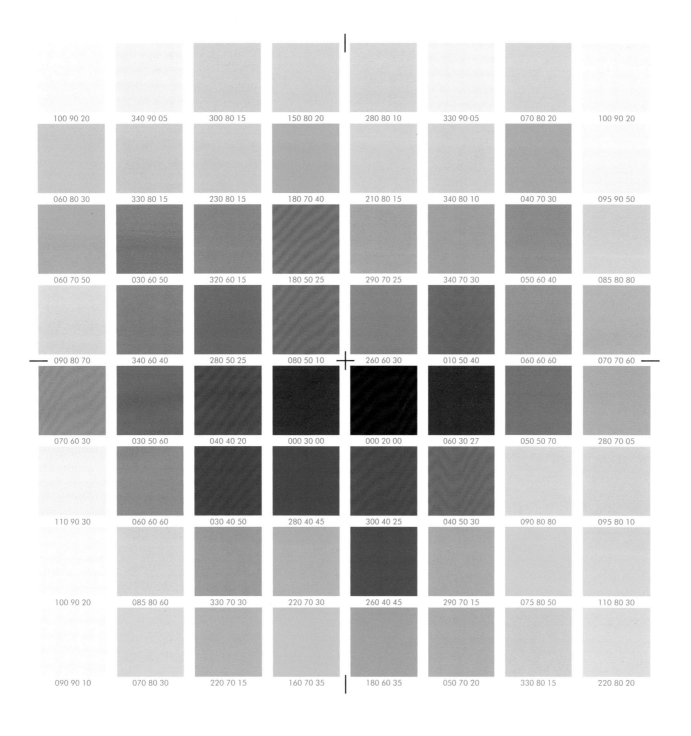

SNUG
舒适温暖

NOSTALGIC
温婉怀旧

FAIRYTALE-LIKE
COSY
SNUG
NOSTALGIC

恍如童话
亲密无间
舒适温暖
温婉怀旧

Whenever the colours reflect a high feel-good factor, they are predominantly warm-toned, mostly unmixed and sensually, animatingly beautiful. The impressions made by the four colour ranges are similar because they are unified by a delicate, cautiously soft and mild foundation. The colours of the "snug" and the "nostalgic" vary regarding their degree of warmth: what we understand by "snug" is frequently reddish, whereas "nostalgic" is interpreted with deeper tones and with brown and grey-black colours.

Nonetheless, when underlining out fields covered by single terms, their independence catches one's eye because they are nuanced more brightly or darkly, more purely or cloudily. The profiling of a type of term frequently shows, especially when repeated, a very finely adjusted result.

能够给人极高舒适度的色彩，必然是温暖、纯粹、富有生气的。这四种用色风格带给人同样的舒适感受，因为它们都有着细腻、柔软、温和的基色。"舒适温暖"与"温婉怀旧"的差异在于温暖的程度：前者多用到红色系，后者则多以更深的棕色和灰黑色诠释主题。

细细对比这四种风格独特的色彩选择，会发现它们或明或暗、或纯粹或杂糅的微妙差异。其中任何一种风格，都需要反复剖析，才能掌握精髓，使之得到恰到好处的运用。

SNUG

舒适温暖

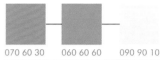

070 60 30　060 60 60　090 90 10

The details are as important as the big picture. Radiating snugness requires warm shades of gold, tea and mint, which, regardless of whether they are viewed in clear light, bright sunlight or by a flickering open fire, have the ability to permit complete relaxation. Every detail must be right, regardless whether it is the brass rails, the picture next to the headboard of the bed or the sand-coloured curtains.

The essential elements when designing are successful combinations. Rooms are nothing other than landscapes in the smallest of spaces. Both on the large and the small scale, they must offer visual axes of stimulation and easing, of animation and quiet. Hotel rooms in particular must provide a microcosm of snugness in a fairly confined space and yet combine cosy closeness with a far-reaching perspective.

090 90 10　070 60 30　060 60 60

An upholstered and leather-covered head section of a bed suggests warmth and additionally generates feelings of security, of natural quality and melodious acoustics. In the bathroom, large mirrors are the most important component; they also enlarge the room and give travellers the opportunity to look at themselves in more detail by affording them a little more leisure than usual ...

细节与整体效果同等重要。"舒适温暖"的氛围要求有温暖的金色、茶色、薄荷色，这些暖色无论是在明媚阳光还是熹微火光下，都能让人感到彻底的放松。无论是黄铜栏杆、床头画框还是沙色窗帘，每一处细节都经过了精心的安排。

成功的设计，基于成功的搭配与组合。房间无异于一处处最小的空间景观。无论房间大或小，都必须有使人轻松、平静的视觉轴线。酒店客房尤其需要在受限的空间中营造自在的氛围，既有室内的亲密舒适之感，又有极目远眺的户外景致。

090 90 10　100 90 20　070 80 30

090 90 10   100 90 20   070 60 30   070 80 30   085 80 60

Many of our male friends talk about it. Exhibiting a bit more vanity may be counted as one of the venial sins, but the result also profits female partners or friends.

床头覆盖有软垫和皮革，天然的质地既舒适温馨，又给人足够的安全感，并能减少噪声，改善睡眠。浴室的大镜面值得一提，能够扩大空间感，还给了顾客们尽情与镜中的自我对视的机会，其从中可找到别样乐趣。

这一空间风格赢得了多数男性同胞的赞赏，稍显浮夸却无可厚非，还俘获了不少女性朋友的芳心。

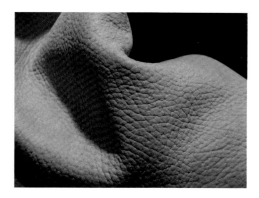

090 90 10   070 60 30   000 30 00   040 40 20

090 90 10   070 60 30   060 60 60   040 40 20

040 40 20	070 80 30

085 80 60

070 60 30

060 60 60

040 40 20

/ Chapter 5

HOTELS FOR CONNOISSEURS AND BOHEMIANS: FROM WEIRD TO DISCREET

Vicki Baum wrote the famous novel "Grand Hotel". It was made into a movie with a top-class cast.

Fifty years ago, grand hotels were the meeting places of high society. Their reputations preceded them as stately, international places where dramatic liaisons, impostures and historical events took place when secret service agents met grey eminences. They were places where famous film stars got to know industrial magnates and impresarios met opera divas. In the movies of those days, gentlemen wore pencil moustaches and Borsalino hats, and they were very elegant. The ladies with tight-laced waistlines wore delicate veils in front of their faces and little feather hats. People still travelled on the "RMS Queen Elizabeth" or "RMS Queen Mary" from continent to continent, or by aeroplane with stops in Shannon and Newfoundland.

It is wonderful when hoteliers remember those days and do not consider "weird" or "nonconformist" to be insults, but incentives to re-stage the well-known. At a time when we divide the service world into areas of expertise, judge people to be "competent" or "incompetent", and become set on the idea that competence, in addition to sustainability and innovation, is one of the main markers of competence, we need not be at all surprised that the struggle to prove competence, as a sign of being "better" and "more diligent", starts to bore us.

Hotels with a heart and soul, in touch with history or the future, which are "chic" or "unusual" or "iconic", but also "turbulent" and "eccentric": that is where I want to be.

Every now and then, one still encounters connoisseurs' options. These come from the heart, intuition and experience; they help to make a decision as to where I want to rest my weary head and where I will refresh my hungry and thirsty body. They are more trustworthy than any tourist guidebook or glossy magazine.

The pages that follow will give examples of what hotels with that special image might look like, and which forms, products and interiors are crucial. We will see that accentuations can be just as conducive in creating an "inspiring and exciting" design, and that the "witty", the "charming" and the very "discreet" has a stimulating effect.

The work of designers always revolves around the same question: "How can one develop the sense of being touched?" Whenever colours, surfaces, shapes, structures, dimensions, light and shadow complement each other and harmonise, then we feel at ease.

鉴赏家与不羁者的酒店：
从捉摸不定到低调慎重

Vicki Baum写过一部知名小说《大饭店》。小说后被改编为同名电影，由一流的演员出演。

五十年前，大饭店是上流社会的专属场所。豪华的大饭店象征着高贵、庄重与国际化，但在光鲜之下，它也见证了充满欺骗、背叛与攻心计的"地下活动"。大饭店还是大明星光顾流连的去处，电影明星会在这里约见片商，剧场老板也在这里挑选女伶。电影镜头下的那个年代，绅士们喜欢留着一字胡，头戴Borsalino帽，品位不凡。女士们则追求束腰之美，还有面纱帽和羽毛头饰。若要出远门，他们会坐上皇家邮轮"伊丽莎白女王号"或"玛丽王后号"，甚至搭乘飞机来一次奢侈的空中之旅。

但愿亲历过那个年代的酒店老板们，不会介意我用"稀奇古怪"或"不入主流"这类词形容他们的酒店，因为我绝没有不敬之意，而是深深留恋。身处今天这个时代，我们已经将专业性作为服务好坏的首要标准，评判一个人时也只认"胜任"或"不胜任"，并理所当然地将胜任力连同忍耐力、创新力视作其胜任与否的主要标志。不过与此同时，不断追求"更好"和"更勤奋"以证明其胜任力，此类励志套路已经开始令我们心生厌烦。

我理想中的酒店，是有精神和灵魂的，能承载历史和未来，它们或"风格别致"或"与众不同"或"特色鲜明"，同时又"愤世嫉俗"且"离经叛道"。时不时，我们还能找到一两处能够俘获鉴赏家芳心的酒店。鉴赏家们全凭心做出选择，跟着感觉走，他们让我意识到，应该选择能让自己的疲惫身心获得充分慰藉的酒店。这一点比任何旅游指南或杂志的推荐都更可靠。

接下来的内容将带你了解特色酒店，阐释它们的外观、室内布置及装饰。我们会发现，设计中的"强调"可以产生令人惊喜的效果，"诙谐活泼"或"低调谨慎"的设计更能增强这种效果。

设计师的工作往往围绕同一个问题："如何触动人心？"色彩、质地、形状、结构、尺寸、光影若能和谐互补，便能给人惬意之感。

320	weird - nonconformist - stark - bold	捉摸不定 – 不入主流 – 了无修饰 – 胆大妄为
332	bohemian - sensual - intoxicating - exaggerated	放荡不羁 – 感官享受 – 令人陶醉 – 潮流浮夸
344	chic - colourful - unusual - spectacular	潇洒别致 – 五彩缤纷 – 与众不同 – 叹为观止
360	turbulent - richly accented - revolutionary - extravagant	汹涌澎湃 – 浓墨重彩 – 颠覆传统 – 肆意放纵
372	iconic - futuristic - eccentric - progressive	标志意义 – 未来主义 – 离经叛道 – 激进革新
386	witty - charming - plain - discreet	妙趣横生 – 妩媚多姿 – 平淡原味 – 低调慎重

/ Chapter 5

WEIRD

The pie chart illustrates that the imagination connects the "weird" with scores of mystical lilac, viola and purple-containing shades.

捉摸不定

圆形图标表明，"稀奇古怪"风格往往青睐神秘的紫色系，包括丁香紫、紫罗兰等含紫色调。

The bar structure suggests that the weird does not work together with the horizontal.

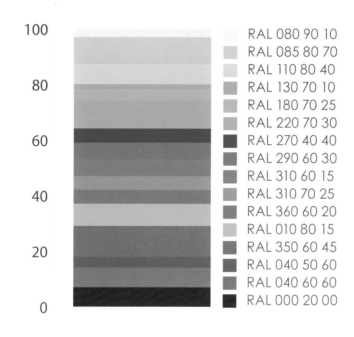

RAL 080 90 10
RAL 085 80 70
RAL 110 80 40
RAL 130 70 10
RAL 180 70 25
RAL 220 70 30
RAL 270 40 40
RAL 290 60 30
RAL 310 60 15
RAL 310 70 25
RAL 360 60 20
RAL 010 80 15
RAL 350 60 45
RAL 040 50 60
RAL 040 60 60
RAL 000 20 00

条形图表明，"稀奇古怪"风格较难与中规中矩的横向线搭配。

More than 50% of slants have a preferred direction originating from our culture. We are often trained as right-handers – therefore we prefer to draw diagonals from bottom left to top right.

受日常文化的影响，超过50%的斜线有着优先定向。比如我们习惯用右手，所以喜欢从左下到右上画对角线。

WEIRD 捉摸不定

Slants are quirky: 25 European examples from bottom left to top right, two the American way from top left to bottom right and one crossed-over. Notice the way that German and American men wear ties. Bush's stripes slant the other way than Schröder's.

古怪的斜线：25个欧洲图样，从左下到右上；3个美式图样，其中2个从左上到右下，另一个反线交叉。你可以留心观察德国和美国男士的领带，比如布什和施罗德，他们的领带或许有着相反的条纹斜线。

/ Chapter 5

NONCONFORMIST

不入主流

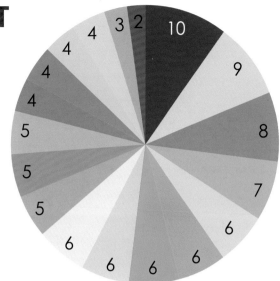

Colours which overall do not seem revolutionary, but generate lascivious, urban associations. The large proportion of black extends over the colouration like an ethereal crape or like a pompous avant-garde demonstration.

"不入主流"风格在用色上并不夸张，却能使人联想起充满魅惑的都市。大比例的黑色像神秘黑纱在着色图上延伸，有着先锋艺术的浮夸色彩。

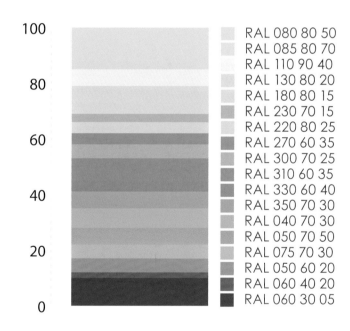

The stripes have a stormy paleness and greyness just as if the arctic cold had transferred its own colour potential to Sicily: a climatic, nonconformist masterpiece.

条形图呈现出大范围的浅色和灰色，好似北极冰寒的冷色移步到了西西里，构成一幅不入主流的气候画作。

RAL 080 80 50
RAL 085 80 70
RAL 110 90 40
RAL 130 80 20
RAL 180 80 15
RAL 230 70 15
RAL 220 80 25
RAL 270 60 35
RAL 300 70 25
RAL 310 60 35
RAL 330 60 40
RAL 350 70 30
RAL 040 70 30
RAL 050 70 50
RAL 075 70 30
RAL 050 60 20
RAL 060 40 20
RAL 060 30 05

The imaginative test participants have found a new home in nonconformism as "hardliners", "neo-cubists" and "late pointillists". They have introduced an air of suspense.

富有想象力的尝试者们，为"不入主流"主义找到新成员，包括"强硬派"、"新立体派"和"后彩派"，为其注入了一种悬疑的氛围。

NONCONFORMIST 不入主流

It is the mixture that matters: nonconformism reveals differentiated and individualised characteristics. Sometimes they manifest in woody cherry and rosewood hues, then in varied lilac-, violet-, royal-blue nuances or plushy fin-de-siècle outfits: in linden green and brass gleam.

重在混合："不入主流"主义有着差异化与个性化的特征，时而呈现樱桃木和花梨木的色调，时而呈现丁香紫、紫罗兰和皇家蓝的色调，或被赋予20世纪末的流行色——椴绿和黄铜色泽。

/ Chapter 5

STARK

了无修饰

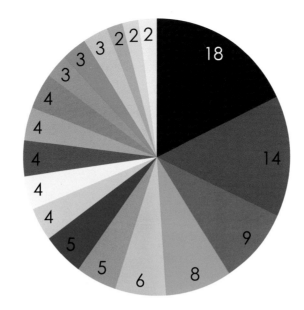

The chart shows high contrasts with shining crayon colours. Bright and dark are used for interpretation in the same way.

饼形图呈现了多种具有高对比度的亮色。暗色与亮色共同诠释了该风格。

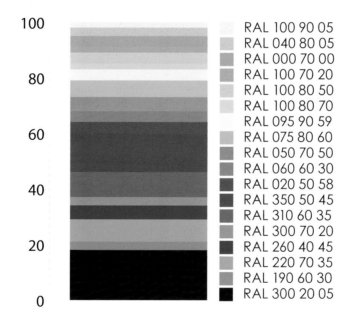

RAL 100 90 05
RAL 040 80 05
RAL 000 70 00
RAL 100 70 20
RAL 100 80 50
RAL 100 80 70
RAL 095 90 59
RAL 075 80 60
RAL 050 70 50
RAL 060 60 30
RAL 020 50 58
RAL 350 50 45
RAL 310 60 35
RAL 300 70 20
RAL 260 40 45
RAL 220 70 35
RAL 190 60 30
RAL 300 20 05

A vigorous sequence with a stable foundation and animated colours has been created. "Stark" is also beautiful, especially for "purists of clarity and conciseness".

表中色彩以黑色为底，由暗至亮平稳过渡。"了无修饰"风格充满魅力，更是"追求明晰扼要"的纯粹主义者的最爱。

Sharp angled, dynamic shapes, mostly with black contours or only hard, thick lines which could be the signatures of berserkers.

锐利的边角、动态的形状、黑色的轮廓、粗硬的线条，种种特征神似北欧传说中的狂暴战士。

STARK 了无修饰

The sharp-angled and the sharp-coloured confirms the stark effect. The experience content is enormous and, strangely enough, ultra-modern, partly ordered but also chaotic, never stolid, almost always dynamic and slender.

尖角与艳色造就了"了无修饰"的设计效果。这一风格给人超强的感官体验，不寻常、后现代、半有序半混乱，始终动感十足而不停滞。

/ Chapter 5

BOLD

胆大妄为

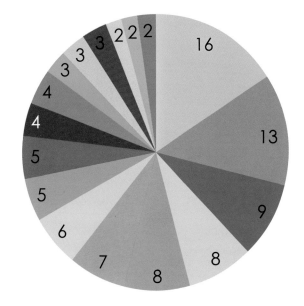

In comparison with the brightly coloured painted fields, the ordered segments are almost harmless. In the appropriate selection, the results become cheekier.

与大面积的明亮色块相比，小比例均匀分布的色样可谓绝对保守。二者调和互补，效果惊人。

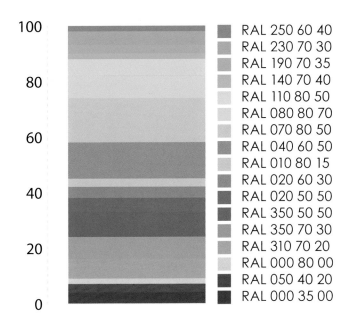

A colour sequence as if designed in Villa Villekulla. A succession of childish, provocative, lovingly playful naughtinesses tasting of sweets.

一系列活泼可爱、充满童趣的颜色，让人仿佛置身于维洛古拉游乐场。

RAL 250 60 40
RAL 230 70 30
RAL 190 70 35
RAL 140 70 40
RAL 110 80 50
RAL 080 80 70
RAL 070 80 50
RAL 040 60 50
RAL 010 80 15
RAL 020 60 30
RAL 020 50 50
RAL 350 50 50
RAL 350 70 30
RAL 310 70 20
RAL 000 80 00
RAL 050 40 20
RAL 000 35 00

Many circular areas, striped patterns, spikes, and little dot patterns. Symbols are rarely as cheeky as colours.

大量使用了圆形、条纹、穗状和小圆点。如此看来，形状并没有颜色那般"胆大妄为"。

BOLD　　　　　　　　　　　　　　　　胆大妄为

The shades are delightful. Promotionally effective, irresistible, not quite painless and rated PG. What bliss for many to wake up in such a room and nibble the deliciously sweet fruit salad.

色调欢快，带来生动迷人、让人心醉神迷的视觉效果。想象一下，在如此甜蜜的小屋中醒来，享用香甜的水果沙拉，该是何等幸福。

/ Chapter 5

WEIRD
捉摸不定

NONCONFORMIST
不入主流

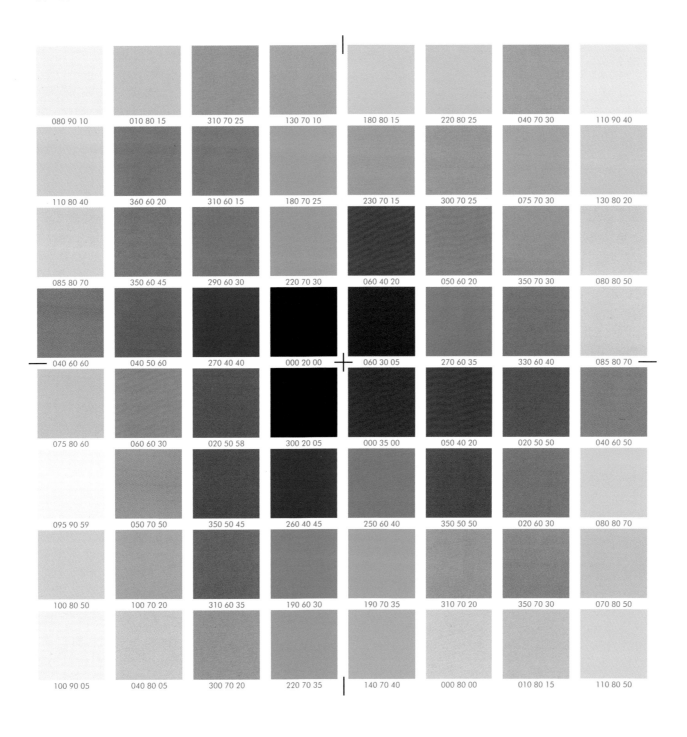

STARK
了无修饰

BOLD
胆大妄为

WEIRD
NONCONFORMIST
STARK
BOLD

捉摸不定
不入主流
了无修饰
胆大妄为

It is obvious at first glance that these are outsider shades. However, they also form part of our design pool and are necessary for the clarification and contrasting of interior design. If they do not serve to mark or signal factual information or aesthetic emphases, they can also be used to ironise small-scale details, or can be activated to add pointed emphases to brand-related attributes.

Equipping a whole hotel in nonconformist colours with stark choices can be successful in a sensational way. Conference rooms, lobbies and bars, too, can be designed with cheeky, quirky and evocative attributes. Big and cosmopolitan cities with esprit can count on guests and companies that have turned away from established conventions because they consider nothing to be less progressive than delighting the world and tomorrow's customers with yesterday's quotations.

乍一看，这些风格貌似只有圈外人才会尝试。其实不然，它们也应纳入我们的备选中，必要时能提升酒店的室内设计。除了突显导视信息或增强审美效果，它们还能升华小范围的设计细节，也能强化品牌相关的设计重点。

如果整个酒店选用"不入主流"的色彩和"了无修饰"的风格，有可能会得到前所未有的惊喜。会议室、大堂和酒吧，也可以尝试这些肆意、古怪而令人回味的元素。如果酒店位于国际化大都市，那么你大可放心顾客的包容力，因为他们一向拒绝既定惯例，绝不会为"陈词滥调"所取悦。

NONCONFORMIST 不入主流

Materiality and illusion complement each other in an exciting way. It is a good thing when the light colour can be changed to create the atmosphere one currently desires: different, funky, or simultaneously quirky and gentle. Views need not always be stable. Changing pictures, and oscillating and changing visual events are welcome variations for those who are looking for new impressions. Between spaceship and pleasure boat, between Captain Nemo and Capitano Esuberanza.

Walls designed with 3D effects and psychedelic roller-coaster patterns: that is how to create a hotel room that every courageous, curious and futuristic-minded spirit would like to try out. The bed, too, with its soft and adventurous silhouette, is worthy of a prize. The carpet decorated with black waves, curves and swinging lines skilfully picks up the love of illustration that characterises the vivid wall backdrops.

060 30 05　050 70 50　310 60 35　270 60 35

现实与幻想互相衬托，碰撞出美妙的火花。灯光变幻出不同颜色，营造出不一样的氛围：可以是另类或时髦，也可以兼得古怪与得体。空间视图不必保持一成不变。变换的图像和回转变动的视觉元素，都能给人留下全新的印象。这种体验犹如坐上飞船或游艇，开启"海底两万里"的奇遇。

墙面设计呈现3D效果，流畅的线条让人联想到过山车：这个酒店房间充满了大胆新奇、未来感十足的设计元素。就连床也因为有了神秘莫测的轮廓而加分不少。地毯展示着黑色的波浪曲线，与动感的墙面背景相呼应。

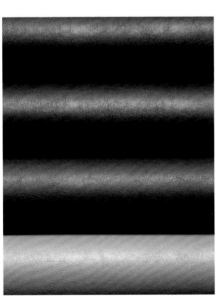

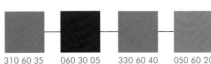

310 60 35　060 30 05　330 60 40　050 60 20

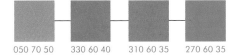

NONCONFORMIST

/ Chapter 5

BOHEMIAN

放荡不羁

The interpreters have made their own sense of it. The delicate hues are almost completely absent. The extraordinary combinations of mud, fog, night, and festive colours define the representations.

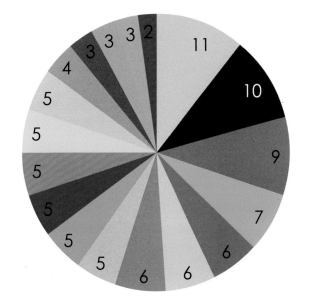

饼形图已经说明一切。甜美的色调几乎不见，取而代之的是泥、雾、夜混合出的色彩，以及具有代表性的喜庆颜色。

The unusual, existentialist side, the expression of which is characterised by violet and purple nuances, is in contrast with the deeper pathetic shades.

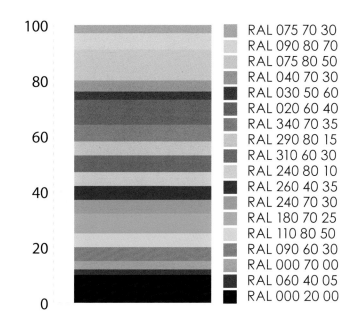

RAL 075 70 30
RAL 090 80 70
RAL 075 80 50
RAL 040 70 30
RAL 030 50 60
RAL 020 60 40
RAL 340 70 35
RAL 290 80 15
RAL 310 60 30
RAL 240 80 10
RAL 260 40 35
RAL 240 70 30
RAL 180 70 25
RAL 110 80 50
RAL 090 60 30
RAL 000 70 00
RAL 060 40 05
RAL 000 20 00

"放荡不羁"风格中不同寻常、具有存在主义意味的一面，体现于紫罗兰色及其他紫色，与更为沉重的深色系形成对比。

The symbols are distinctive, slanting, dynamic, as if carved or hewn with a hammer.

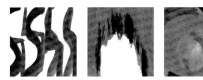

这些符号个性十足，呈现出倾斜而动态的效果，就像是用锤子凿刻出来的。

BOHEMIAN　　　　　　　　　　　　放荡不羁

The bohemian is not for the faint-hearted. Not suited for overly shallow operettas. Better suited for mini dramas. Bohemia never trusted in the beautiful shades. In the colour fields, this is illustrated by the wrong shades in a declamatory way.

"放荡不羁"风格不是胆小者的选择，也不适合轻歌剧式场景，但能营造出迷你剧式的效果。这一风格从不追求漂亮的色调，却以事实胜于雄辩的方式证明其独特的用色之道。

/ Chapter 5

SENSUAL

感官享受

The pie chart resembles its culinary equivalent, which was conjured up out of raspberries, strawberries and a few appealing ingredients by a colour-loving confectioner.

这一圆形图标好似厨房烹饪，搅拌着令人垂涎的覆盆子和草莓，再洒上诱人的配料，堪称甜点师手中的色彩大作。

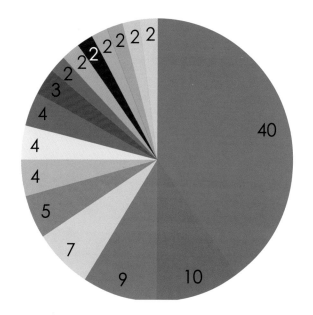

Who is surprised? Vibrant red determines the picture. Some accompanying shades applaud the sensual in a rather tongue-in-cheek way.

谁会惊讶呢？动感的红色成为主宰。其他色彩通通沦为配角。

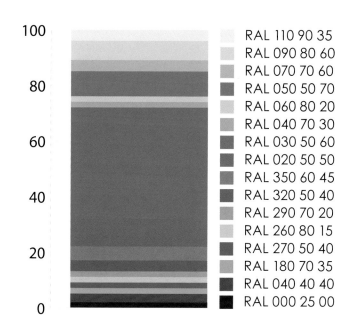

RAL 110 90 35
RAL 090 80 60
RAL 070 70 60
RAL 050 50 70
RAL 060 80 20
RAL 040 70 30
RAL 030 50 60
RAL 020 50 50
RAL 350 60 45
RAL 320 50 40
RAL 290 70 20
RAL 260 80 15
RAL 270 50 40
RAL 180 70 35
RAL 040 40 40
RAL 000 25 00

Circles, so it seems, suffice to give the theme a symbol.

不难发现，圆形已然成为这一风格的符号象征。

SENSUAL   感官享受

The prevalence is obvious. No pleasure garden without red roses, no temple of abundance without red. Only amusements wear the purely colourful, as the colour pictures of funfairs, eve-of-wedding parties and company parties prove.

红色成为流行色理所应当。红玫瑰为花园添姿加彩,红色也为庙宇营造出庄严气氛。游乐场、婚礼前夜派对、公司聚会等娱乐活动,更少不了刮起红色着装的旋风。

/ Chapter 5

INTOXICATING

令人陶醉

The psychedelically charged, substantial shades with a large proportion of deep red determine the range.

这些颜色隐藏着催人产生幻觉的潜质，深红色占据了最大的比例。

The arc of suspense ranges from cognac-orange to red, garish red, scarlet through to light blue. The ecstatic needs neither yellow nor green, nor brown; the rational blue, however, is far from absent.

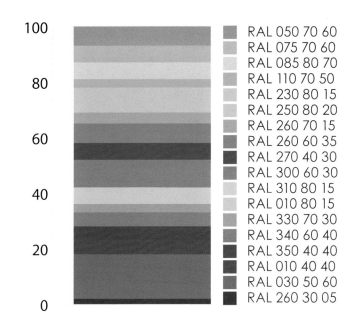

	RAL 050 70 60
	RAL 075 70 60
	RAL 085 80 70
	RAL 110 70 50
	RAL 230 80 15
	RAL 250 80 20
	RAL 260 70 15
	RAL 260 60 35
	RAL 270 40 30
	RAL 300 60 30
	RAL 310 80 15
	RAL 010 80 15
	RAL 330 70 30
	RAL 340 60 40
	RAL 350 40 40
	RAL 010 40 40
	RAL 030 50 60
	RAL 260 30 05

条形色由柳橙、干邑色渐变为鲜红、猩红色，浅蓝充当过渡色彩。想要营造迷离而引人入胜的氛围，应该选择的不是绿或棕色，而是理性的蓝色。

Plenty of slants, spiral shapes and, in one case, also something explosive. The chaotic and the wavy are part of the symbolic equipment.

大量的斜线和螺旋状图案，迸发出强劲的表现力。混沌朦胧、起伏不定是这一风格的典型特征。

INTOXICATING   令人陶醉

Colours of New Year's Eve parties and film balls from all over the world. No rosé, yellow or bright green dares to enter the colour range. Shades of classic decadence and baroque plenitude of splendour combine as a bacchanalian jamboree.

图中色彩让人想起了跨年派对或庆祝舞会，具有节日氛围。玫瑰红、黄色和鲜绿色没有在色域中出现。复古颓废与巴洛克的辉煌色调相融合，色彩的狂欢就此诞生。

/ Chapter 5

EXAGGERATED

潮流浮夸

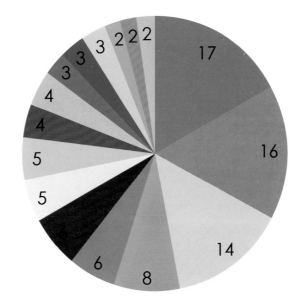

The circular chart proves that the loud always requires a high level of colourfulness. Bright-dark contrasts increase their effective quality.

色盘表明，要想吸睛，就少不了缤纷绚丽的色彩。明与暗的色彩对比，更加强了这种效果。

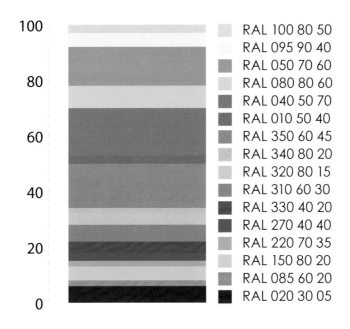

RAL 100 80 50
RAL 095 90 40
RAL 050 70 60
RAL 080 80 60
RAL 040 50 70
RAL 010 50 40
RAL 350 60 45
RAL 340 80 20
RAL 320 80 15
RAL 310 60 30
RAL 330 40 20
RAL 270 40 40
RAL 220 70 35
RAL 150 80 20
RAL 085 60 20
RAL 020 30 05

A rather balanced warm-cold colour field has been produced. The red-inflected shades represent a qualitative majority.

条形色呈现暖色与冷色的和谐过渡。红色占据了很大的比例。

The graphic signals are interpreted with stable, powerful elements and are represented in surface fields.

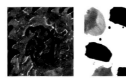

图形标志以稳固而有力的元素为主，常被运用于空间的表面装饰设计中。

EXAGGERATED 潮流浮夸

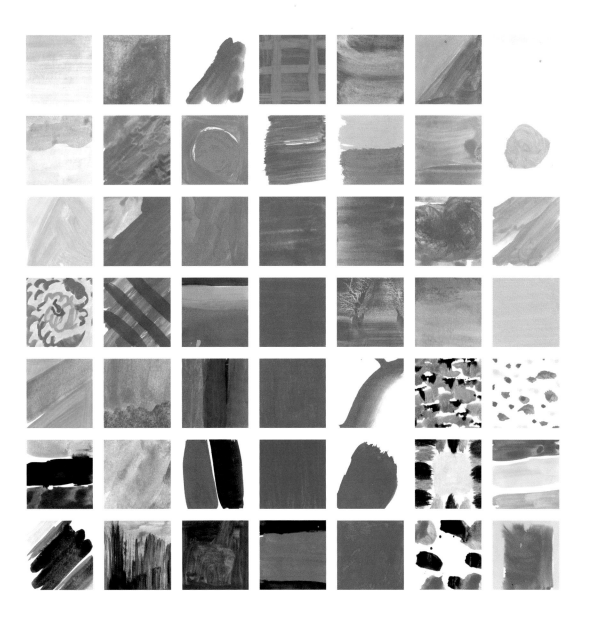

Magnificent contrasts unfold. An extravagant, very exciting mixture develops, consisting of less effective golden brown and aggressive black-magenta.

色块之间同样对比鲜明。浮夸式的色彩混搭令人兴奋，包括相对保守的金棕色和咄咄逼人的黑色与洋红色。

/ Chapter 5

BOHEMIAN
放荡不羁

SENSUAL
感官享受

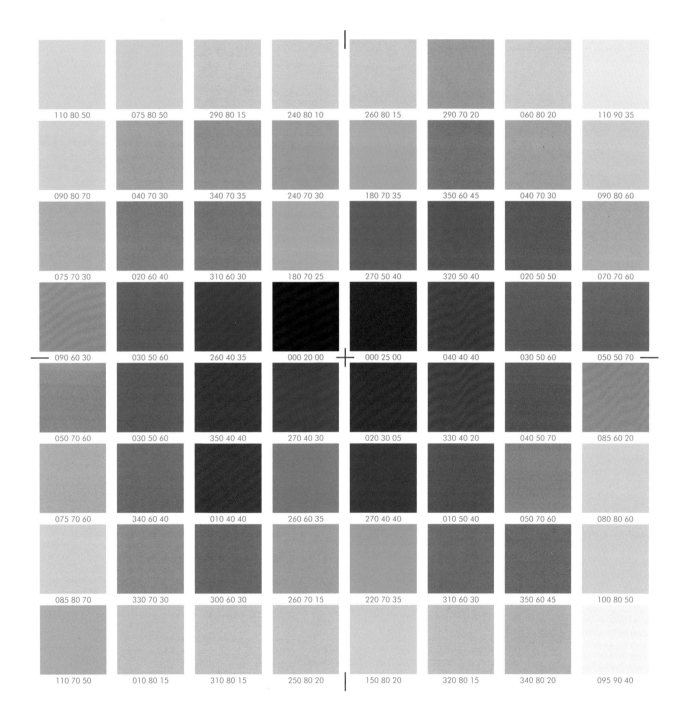

INTOXICATING
令人陶醉

EXAGGERATED
潮流浮夸

BOHEMIAN SENSUAL INTOXICATING EXAGGERATED

放荡不羁
感官享受
令人陶醉
潮流浮夸

Each of the colour fields comprising sixteen squares has a similar dramatic quality. The large number of lascivious pastel hues, which can trace their origins to violet, rose, elder, lavender, clematis and orchid shades, characterises the overall picture. The most striking appeals to the senses emanate from the reddened nuances, followed by blue, and then orange-yellow colourings.

The contents of such colourfulness originally developed outside bourgeois conventions. The bohème, a circle of artist friends, was formed in the first half of the nineteenth century. The lives of bohemians, with their feasts, states of intoxication, conflicts and psychological and existential conditions, were dramatised by Giacomo Puccini in his opera "La Bohème" in 1896. As a design and lifestyle metaphor, it is still quoted as a libidinous, intoxicating and exalted background model.

每一风格各自有16个色块，都能帮助营造戏剧化的氛围。挑动欲望的蜡笔色彩大量出现，让人联想到紫罗兰、玫瑰、接骨木、薰衣草、铁线莲和兰花等植物。而最能刺激感官的色彩，莫过于红色，其次是蓝色和橙黄色。

这些色彩最初由中产阶级的生活习惯中传承而来。波西米亚风格，由一群年轻艺术家在19世纪上半叶创立。波西米亚人的生活方式，包括他们的盛大节日、酒池肉林、冲突矛盾、心理状态以及生存状况，都能在吉亚卡摩·普契尼于1896年创作的歌剧《波西米亚人》中找到原型。作为一种有代表性的设计与生活方式，波西米亚风格依旧有着摄人心魄、令人陶醉的持久魅力。

BOHEMIAN 放荡不羁

One can easily imagine a bohemian soirée taking place in this room. The slightly exaggerated bourgeois style, which consists of black-and-red satiation, was as suitable for the artists and mavericks of bygone eras as it is for those who are in touch with today's zeitgeist. It is wonderful when interiors reveal nuances. It is also wonderful when we, at second or third glance, realise that everything need not be taken too seriously, that rooms can represent serious decoration but also tongue-in-cheek upper-class elegance. Good restaurants always leave enough room for the grand entrance in order to offer all those already present topics of conversation that will last for the rest of the evening and the night.

There is no need for too much pomp and splendour on walls and ceilings here. Only that which remains silent in the face of vitreous glitter is elegant. It is magnificent that there are hotel designs that have only one thing in mind, namely focusing on the human being. That is why big mirrors are a must. They provide us with welcome access to our genetic mirror image, which we either love or hate. Either way, it is familiar.

000 20 00   075 80 50

这个房间就像是要举行一场波西米亚式的社交晚会。略显隆重的中产阶级风格，其中包括饱和的黑与红色调，很能迎合新潮流下依然保持旧派作风的艺术家和标新立异者。室内的精致装饰让人眼前一亮，而再看第二眼或第三眼时你才恍然大悟，其实不必过分斟酌室内细节，因为虽然是用心布置，却也不乏无伤大雅的诙谐元素。好的餐厅设计总会为正大门留出足够的空间，以方便来宾之间三三两两地畅快交流，毕竟社交才是当晚最重要、最耗时的重头戏。

075 70 30   075 80 50   090 60 30

墙壁和天花板无须增加太过华丽的装饰。光彩熠熠的玻璃灯饰就足以增添高雅之感。好的酒店设计只需牢记一点，即以人为本。由此便知，大大的镜子为何必不可少。

000 20 00   060 40 05

大镜子在迎宾通道上迎接着顾客的到来,它们映照出顾客的样貌,不管样貌漂不漂亮,却都是真实的样子,都会给予我们熟悉感。

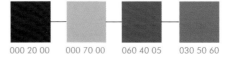

CHIC

潇洒别致

The range of red nuances is more comprehensive than the variety of blue and yellow hues. The design model "chic" is convincing – also for the interior!

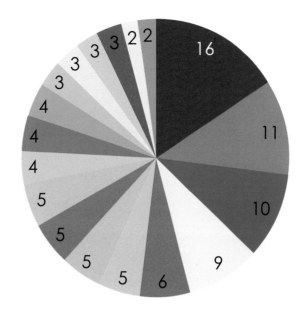

与蓝色和黄色相比,红色系的色样更加广泛。"潇洒别致"的设计总是深得人心,室内设计也不例外。

The bars are a treat for the eyes in every dimension. The gentle, loud and colourful tones confirm a very rewarding wealth of combinations.

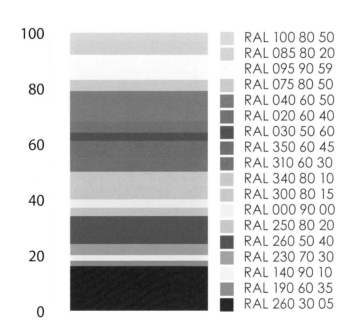

RAL 100 80 50
RAL 085 80 20
RAL 095 90 59
RAL 075 80 50
RAL 040 60 50
RAL 020 60 40
RAL 030 50 60
RAL 350 60 45
RAL 310 60 30
RAL 340 80 10
RAL 300 80 15
RAL 000 90 00
RAL 250 80 20
RAL 260 50 40
RAL 230 70 30
RAL 140 90 10
RAL 190 60 35
RAL 260 30 05

这些条形的色彩式样,每一种都非常养眼。色调有柔和的、强烈的、缤纷的,完美地融合在一起。

"Predominantly robust and angular" is the metaphor of the semiotic meaning.

结实的构造和分明的棱角,是这一风格暗含的符号意义。

CHIC

潇洒别致

Spring-summer fashion shades as if inspired by images from Elle, Harper's Bazaar, Madame and Marie Claire constitute the repertoire of beautiful, elegant, feminine colours. Black represents the top shade of omnipresent chic.

欣赏图中的色彩，仿佛是在翻阅Elle、Harper's Bazaar、Madame、Marie Claire等时尚杂志，在书中捕捉春夏秀场的潮流风向以及美丽、优雅的女装流行色。黑色是"潇洒别致"风格永不过时的经典。

/ Chapter 5

COLOURFUL

五彩缤纷

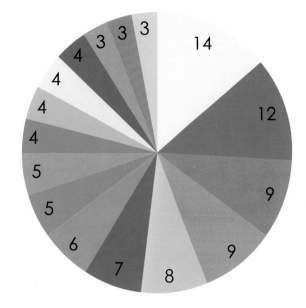

A colour spectrum of great narrative power and gentle urgency. Divided into segments, it is very stimulating and extremely communicative.

整个色谱有着强大的叙述力和温柔、和缓的魅力。单个看来，又能激发人们的兴趣、鼓励交流。

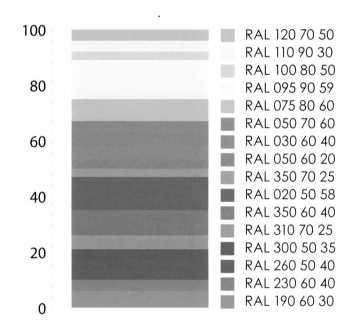

The colour range is beautiful, lovely and graceful. The cool and the warm tones more or less balance each other out.

色彩整体优雅迷人，冷暖色调和谐互补，拿捏得当。

The number of pointillist representations preponderates by far, followed by stripes, amorphies and circles.

图案以点彩派为主导，其次是条纹的、无定形的及圆圈的形状。

COLOURFUL   五彩缤纷

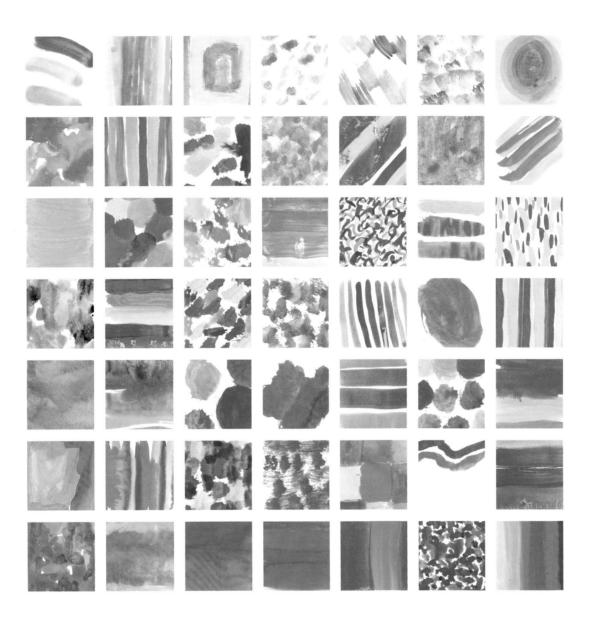

Vernal colours drift towards us. Yearning for childhood, innocence and the blissfulness of romantic roundels and glossy prints arises. Modernistic interpretational ideas are virtually absent. It is interpreted much too archaically and categorically for that.

春天的色彩扑面而来，唤起我们纯真的童年回忆，也让人联想到浪漫的圆舞曲和平滑的版画。当代的诠释手法未被采用，而以直截了当的方式将之呈现。

/ Chapter 5

UNUSUAL

与众不同

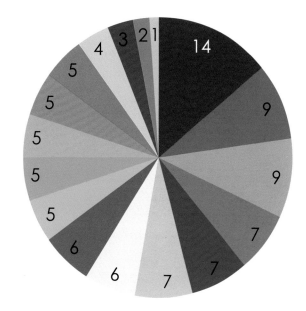

This colour circle also has its exceptional qualities, which are worthy of a wizard and storyteller.

所有颜色有着独一无二的特质，值得细细品味、静静感受。

The colour sequence confirms the inherent narrative quality: it is not real, but fairylike, full of fantasy and poetry.

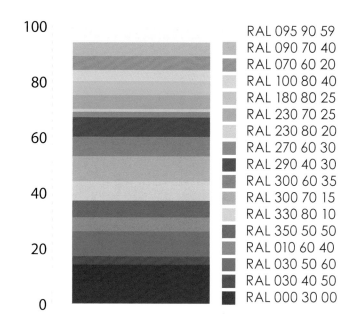

RAL 095 90 59
RAL 090 70 40
RAL 070 60 20
RAL 100 80 40
RAL 180 80 25
RAL 230 70 25
RAL 230 80 20
RAL 270 60 30
RAL 290 40 30
RAL 300 60 35
RAL 300 70 15
RAL 330 80 10
RAL 350 50 50
RAL 010 60 40
RAL 030 50 60
RAL 030 40 50
RAL 000 30 00

这些颜色验证了这一风格与生俱来的叙述品格：它不是真实的，而是如仙境一般，充满诗意与幻想。

Pyramids, diagonals, zigzags, waves, and amorphous shapes belong to the repertoire of symbols.

图形标志包括棱锥、对角、锯齿、波浪和和无定形的图形。

UNUSUAL   与众不同

The unusual releases unknown creations. The search for colours accommodating this adjective has yielded wonderful results. The unusual is what amazes us: mouldy shades of grey, elegant, mystic pink and pure azure colours.

"与众不同"意味着未知的创意。与之对应的色彩呈现出令人惊叹的效果，让我们大吃一惊：包括朦胧的霉灰色、优雅神秘的粉色和纯净无瑕的淡青色。

/ Chapter 5

SPECTACULAR

叹为观止

It is the unusual appearance that is convincing. That is why grey and brown and green do not exists. Black plays the role of the clarifying colour.

不同寻常的外观最具说服力。所以常见的灰色、棕色和绿色并未出现。黑色是其中的主导色。

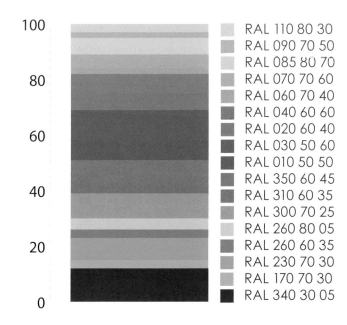

The colour scale emphasises the preciousness of the sequence. It turns every gala dress into a visual pleasure. It reminds us of the Emperor Waltz and gown-filled halls of mirrors underneath chandeliers.

色度表的色彩排列极其讲究，就像一件晚会礼裙带给人的视觉享受。这件色彩之裙让我们联想到皇帝圆舞曲以及华丽吊灯下光滑如镜的舞池。

Colour vignettes, gems and concentric objects, jewellery designs, diamonds, lapis lazuli, rubies, turquoises, and aquamarines.

图形看起来像晕影、钻石、同心纹、天青石靛、绿松石和海蓝宝石。

SPECTACULAR 叹为观止

The spectacular is rooted in this world. The chic, blazing colours indicate the fascination of visual occurrences. Mystification or subtleness are immaterial.

营造"叹为观止"的效果重在用色。别致而闪耀的色彩有着强烈的视觉吸引力。神秘或微妙的元素则可有可无。

/ Chapter 5

CHIC
潇洒别致

COLOURFUL
五彩缤纷

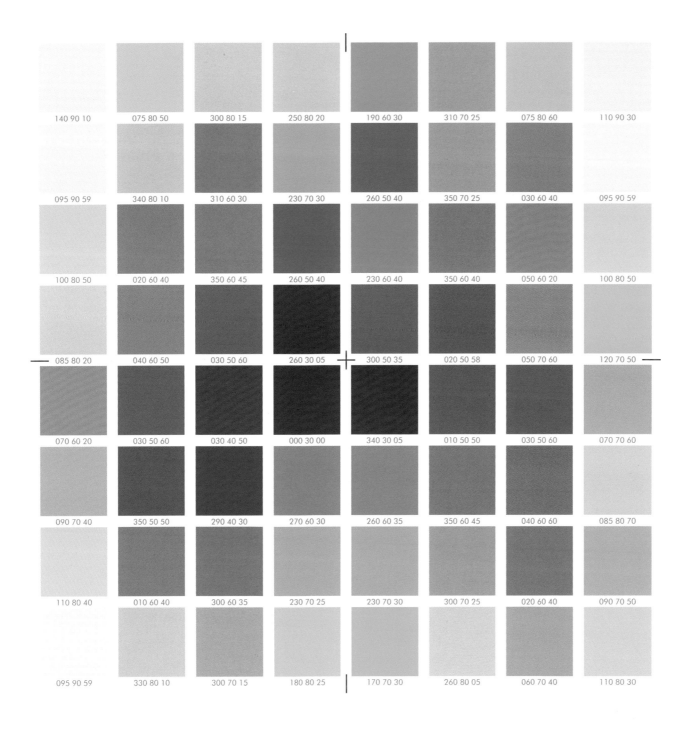

UNUSUAL
与众不同

SPECTACULAR
叹为观止

CHIC
COLOURFUL
UNUSUAL
SPECTACULAR

潇洒别致
五彩缤纷
与众不同
叹为观止

If it is meant to be a truly spectacular, dynamic and stormy affair, then one needs visual treats. The opportunity to stay in such a hotel presents itself very rarely. The colour ranges become more expressive and thus their entertainment value increases. Coloured metals are more in demand than ever. Matt and glossy finishes are applied as dramaturgical tools more frequently in order to create more valuable impressions through surfaces. Experiential values are the driving force behind good entertainment, and they stay in people's memories.

"Chic" and "unusual" are very attractive qualities. The entertainment level approaches the maximum. From our own experience we know that after staying in an El Dorado of colour for two or three days, a stimulating normality arises, and we are afraid that life will be more boring afterwards.

令人叹为观止、如狂风暴雨般震撼人心的空间，犹如一场视觉盛宴。这些色彩给人强烈的视觉冲击，利于提升空间的娱乐价值。有色金属的色彩从未如此抢眼。亚光和有光泽的饰面利于营造戏剧化，使人印象深刻。除了良好的娱乐功能之外，空间也因此更具体验价值，给顾客留下美好的记忆。

"潇洒别致"和"与众不同"是极具吸引力的品质。娱乐功能得到了最大限度的发挥。想象一下，如果你在一座黄金城住上两三天，就会逐渐对金碧辉煌的氛围习以为常，并担心回到普通生活之后会觉得更单调乏味。

CHIC   潇洒别致

Chic and charm form a symbiotic unity. Fashion elements and zeitgeist appeals are as in demand as elegance and high standards. Again and again, this term is assessed in terms of atmosphere, a visually pleasing appearance and taste. With regard to the ambience, values such as order and clarity and tempered joie de vivre are also present. The most striking feature of a "chic interior" is frequently that it does not look draped and improvised but makes do with subtle gestures.

The design of the armchairs is harmonious and ergonomic, bolide-like and comfortable. The button-tufted lines are fluid, as are the lattice fabrics, the artificial light and the sunlight filtered inside.

350 60 45   310 60 30   030 50 60   075 80 50

085 80 20   030 50 60   310 …

The classic plain colours of the fabrics arranged in a fan-like manner, the soft upholstery shades and the elegant beige hue of the leather corner bench stand for charming colouring.

别致与魅力共存，时尚元素与时代思潮兼容，优雅与高标准并重。"潇洒别致"常被用来描述空间环境，意味着赏心悦目的外观与高雅的品位。关于环境，则要求是有序、清晰和温和的，使人尽享生活之乐。"潇洒别致"的室内设计最显著的特征是，不给人皱皱巴巴、临时发挥之感，而是前期早已有了精心酝酿。

085 80 20   030 50 60   260 30 05

扶手椅与空间氛围如此和谐，流线型的设计无比舒适。纽扣凹钉的线条十分流畅，格子织物、人工照明、窗外射进来的阳光也都带来完美享受。

织物为经典的原色，呈扇形布置。软装饰带来温馨之感，皮革转角台是优雅的米色色调，让人着迷。

000 90 00   085 80 20   300 80 15

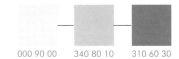
000 90 00   340 80 10   310 60 30

That which is chic is never loud, or garish and ordinary. The chic depends on connoisseurship because its proximity to luxury is understated. By the way, that which is chic is also cosy and always comfortable and light, and does not seek applause or admiration.

Chic rooms" are always human and never superhuman. They meet with our approval or appreciation because they display homely qualities and bring the beautiful into harmony with the factually necessary. If we transfer what we consider chic to a person, we think of quality in taste, without exaggeration, and topped off with a great sense of the moment and esprit.

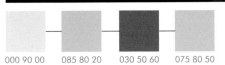

000 90 00   085 80 20   030 50 60   075 80 50

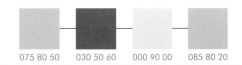

075 80 50   030 50 60   000 90 00   085 80 20

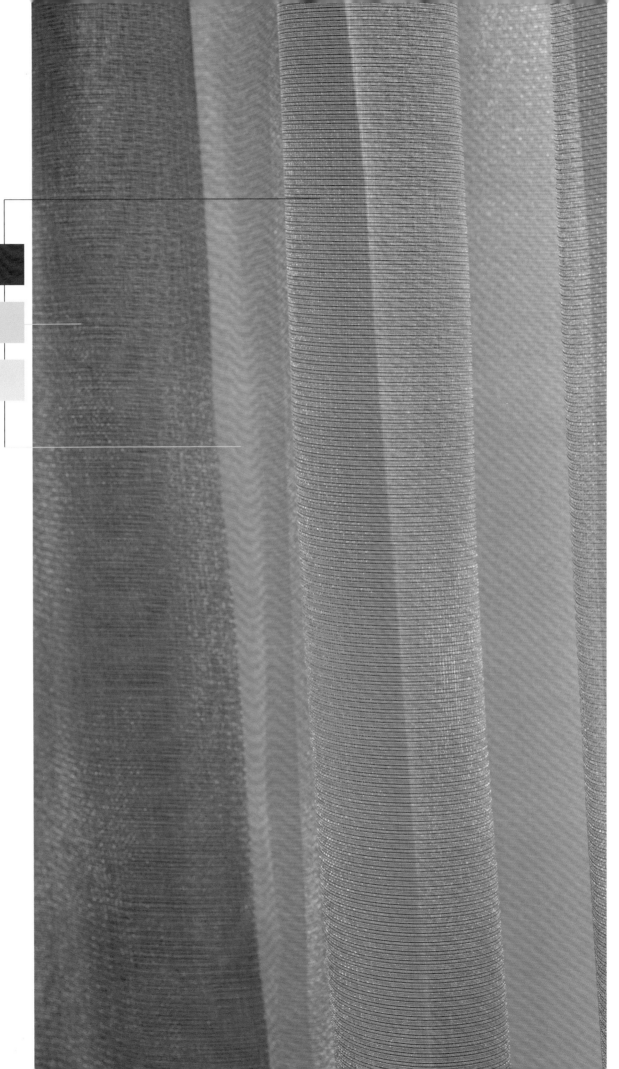

260 30 05

075 80 50

085 80 20

040 60 50   030 50 60

075 80 50　085 80 20　000 90 00

　　"潇洒别致"的风格从来不喧哗，也绝不花哨或平庸。这就要求有精准的鉴赏力，因为它与奢华风格看似很接近。同时，"潇洒别致"也意味着温馨、舒适和柔和的光线，而不会刻意制造亮点吸引人们的注意力。

　　"潇洒别致"的房间从来都是以人为本，而不是"超人"。它们满足我们的实际需求，以朴实的品质与和谐优美的氛围赢得我们的青睐。如果用这个词来形容一个人，也表明他具有不浮夸的本真，有着卓越的心智和品格。

085 80 20　260 30 05

085 80 20

030 50 60

000 90 00

075 80 50

/ Chapter 5

TURBULENT

汹涌澎湃

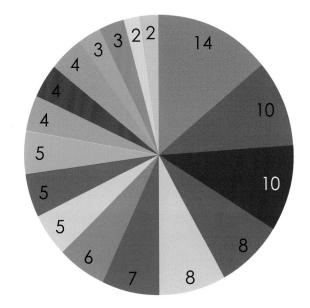

The colour circle proves it: the composition of turbulence is characterised by the two primary colours red and green. Here and there, black is chosen as a competing contrast.

"汹涌澎湃"的用色，以红色和绿色为主色调。其次是黑色，与之形成对比。

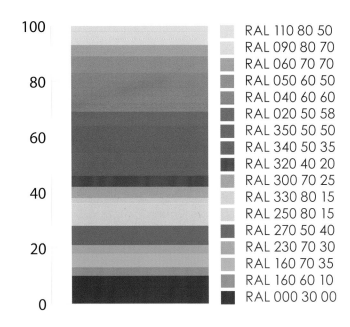

The colour sequence also shows the different brightened up shades of blue and the differentiated nuances of the reddish shades.

RAL 110 80 50
RAL 090 80 70
RAL 060 70 70
RAL 050 60 50
RAL 040 60 60
RAL 020 50 58
RAL 350 50 50
RAL 340 50 35
RAL 320 40 20
RAL 300 70 25
RAL 330 80 15
RAL 250 80 15
RAL 270 50 40
RAL 230 70 30
RAL 160 70 35
RAL 160 60 10
RAL 000 30 00

条形色显示大面积的蓝色系，以及有着不同色差的红色系。

The test participants imagine the symbols of this adjective as being dynamically agitated and hardly subject to an organising principle, except for sequences and hatching-like surfaces.

研究者所想到的图案特征是，动态回旋而无固定规律，除了序列和类似阴影的表面。

TURBULENT

汹涌澎湃

Red and blue are the absolutely approved symbols of the temperamental. Yellow or pink, but also green with pink and every beige or cream shading are despised. The test participants agreed: red and blue colourings are the favourites.

红色与蓝色显然是与这一风格最匹配的色彩。黄色或粉色、绿色与粉色，以及米白色、奶油色，都不推荐尝试。总之，研究参与者一致认为：红色系与蓝色系是最适宜的。

/ Chapter 5

RICHLY ACCENTED

浓墨重彩

Vibrant colours prevail in the pie chart against the brighter and harmless shades. If used, black, red and yellow are given a quantitatively measurable dominance.

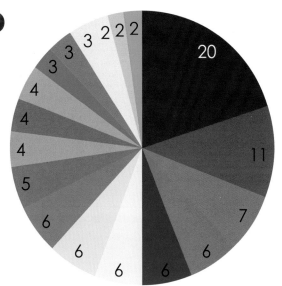

充满活力的色彩在圆形分格图表中占上风，过亮或过于温和的色彩只能退至一角。从数量上看，黑色、红色和黄色都占据着主导地位。

The fiercely loud shades and the bright, clear nuances are the decisive ones. Earth-tinged colourings are almost completely absent in the medley of colours.

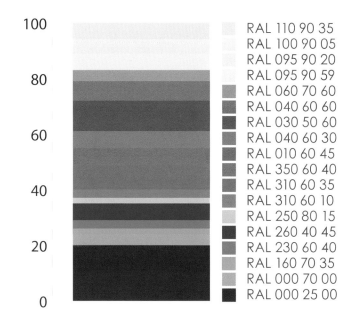

起决定作用的是浓墨重彩和明亮干净的色调。略带泥色的色彩则几乎湮没在这首色彩鸣奏曲中。

The predominantly strict, dark-shaded lines and colour blocks characterise the creative drawing style.

肃穆而有着深色阴影的线条和色块非常显眼，成为这种创作绘画风格的特征。

RICHLY ACCENTED 浓墨重彩

The colours are tinged with black. A host of reddish and pastel tones are also discernible. The drawing style is unanimously given interpretative priority over colour-fulness.

色彩都略带黑色。许多柔和的色调和淡红色调也清晰可辨。此种绘图风格在选色上，都优先选取过于饱满的颜色。

/ Chapter 5

REVOLUTIONARY

颠覆传统

Yellow, green and delicate blue nuances alleviate the sharpness and garnish the term with simple mollifications

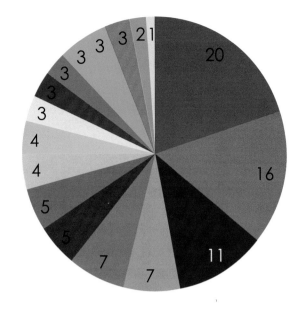

黄色、绿色和浅蓝色减弱了尖锐感，使得圆形分格图看起来较柔和。

The bar chart assumes more bourgeois virtues than expected. Only when related to the content of the signs does the explosive power become more.

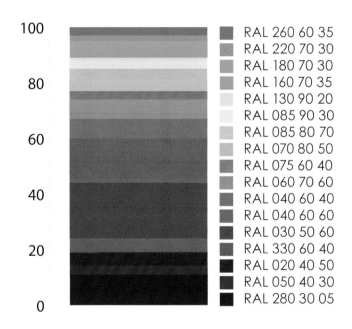

RAL 260 60 35
RAL 220 70 30
RAL 180 70 30
RAL 160 70 35
RAL 130 90 20
RAL 085 90 30
RAL 085 80 70
RAL 070 80 50
RAL 075 60 40
RAL 060 70 60
RAL 040 60 40
RAL 040 60 60
RAL 030 50 60
RAL 330 60 40
RAL 020 40 50
RAL 050 40 30
RAL 280 30 05

条形图看起来比预期的更显中庸务实。只有颠覆传统风格的色彩才有更强的爆发力。

The dynamic of the lines and structures striving from the inside to the outside illustrates the coloured essence.

线条和结构由内至外的动感揭示了这种风格的着色精髓。

REVOLUTIONARY   颠覆传统

Black and red are the colours of buccaneers, of revolution and power, which is enforced by all available means. If we refer to something as revolutionary, we always imagine the same colour pictures. That is barely subversive or new, and therefore an evolutionary colour standard.

　　黑色和红色作为海盗使用的色彩，象征着利用一切可行手段强制推行的变革和力量。如果我们说某物为革命性的，我们总会想到这样的彩色画面。这几乎是颠覆性的或全新的，成为升级了的色彩标准。

/ Chapter 5

EXTRAVAGANT

肆意放纵

A true cake fight has started. One colour goes for another one. This is the recipe for the typically extravagant tones, and they are a spitting image of sporty cyclist's gear on Sundays.

"蛋糕"争夺战真正开始了。一种颜色在追逐另一种颜色。这是肆意放纵色彩风格的典型配色方法，它们看起来和周日骑行者的自行车的齿轮一模一样。

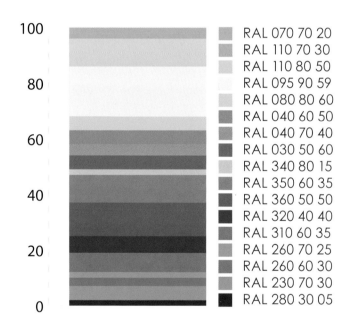

RAL 070 70 20
RAL 110 70 30
RAL 110 80 50
RAL 095 90 59
RAL 080 80 60
RAL 040 60 50
RAL 040 70 40
RAL 030 50 60
RAL 340 80 15
RAL 350 60 35
RAL 360 50 50
RAL 320 40 40
RAL 310 60 35
RAL 260 70 25
RAL 260 60 30
RAL 230 70 30
RAL 280 30 05

Effects of this range are characteristic for a majority of colour worlds from the Catch & Carry segment: so loud they might cause visual tinnitus.

色彩界的大多数争抢会带来图表所示的效果，它们如此花哨，可能会引起视觉不适。

A complete mess of subdued shapes in accordance with colour disharmonies. Conflicts with the shrill are always to the disadvantage of the less shrill.

色彩不和谐使得图形也失去了形状，完全是一团糟。和强势的色彩发生冲突时，稍弱的色彩往往会败下阵来。

EXTRAVAGANT 肆意放纵

The colours' behaviour is biting, cheeky and bickering. The combinations and not the individual shades are the ones that make us shudder.

图中的色彩仿佛在撕咬、争吵。并非单个的色调，而是这种结合让我们不寒而栗。

/ Chapter 5

TURBULENT
汹涌澎湃

RICHLY ACCENTED
浓墨重彩

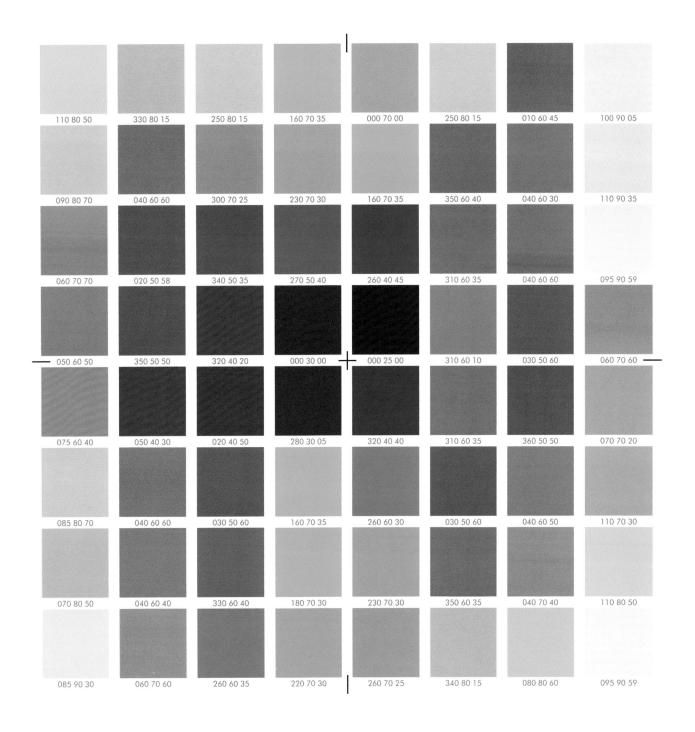

REVOLUTIONARY
颠覆传统

EXTRAVAGANT
肆意放纵

TURBULENT
RICHLY ACCENTED
REVOLUTIONARY
EXTRAVAGANT

汹涌澎湃
浓墨重彩
颠覆传统
肆意放纵

The shades do not even look too colourful to us: they have an edge and illustrate something that intends to grab our attention. Colours have several talents to fulfil: they range from ingratiation and sweetening to shrill signals and visual syncopes. We also know that pure egg-yolk yellow juxtaposed with blue-violet, or bright green juxtaposed with clean purple, represents both a catastrophe and a visual treat. The decisive features are the nuances of the colour content, the degree of glossiness and the quantitative framework, as well as the applied design background.

The colour brown appears only in conjunction with the adjective "revolutionary". Black and vibrant green are missing when it comes to "extravagant"; in the "turbulent" section it looks as if the shades are all over the place; and "accentuated" features contrasting colourful and bright-dark values. In the planning phase of our research, we also had a keen interest in the "extreme" typecasting of colour. Its aim was to animate the participants in the study more intensively using semantic differentials such as "historic" and "futuristic" or "festive" and "plain"; and to keep the senses engaged for focussing.

这些色调看起来并不丰富多彩,它们有些优势,试图抓住我们的注意力。有些色彩有着从逢迎取悦的色彩转变为刺眼的信号和视觉杀手的天分。正如我们所了解的,纯蛋黄色和紫罗兰色、翠绿色和紫色的并置,可能因违和而成为灾难,也可能因巧妙的撞色而奉上一场视觉盛宴。起决定作用的是色彩内涵的细微差别、光泽度和所占比重,以及所应用的设计背景。

褐色只与"革命"有关;黑色和鲜绿色不在"肆意放纵"的讨论范围内;而在"汹涌澎湃"一节中,所有色调看起来都略显动荡而不在常态;鲜艳明亮和黑暗的色彩反差则非常"突出"。在研究规划阶段,我们也对色彩 "极端"的角色定位产生了浓厚的兴趣。其目的在于鼓励参与者们在研究中更集中关注如"历史的"和"未来主义的"或"节日的"和"平实的"等词的语义差异,同时随时保持清醒的头脑。

RICHLY ACCENTED 浓墨重彩

Adding accentuations is part of designers' work, as is the introduction of harmonious references into scenic ambiences. Architectural treats like a circular porthole that casts the light that enters onto the comfortable double bed are frequently used. Here, the circular windows appear like distinctive (saucer) eyes peering out of the room, constituting a sentimental, conspicuously cosy element. The room gains authenticity because it disrupts this first impression with numerous contrasts by using distinctive shades like red and anthracite.

Light immerses the bright walls in warm shades as if lit by sunlight. Most people perceive roof slopes as pleasant, because their slants and tapering heights provoke feelings of safety and comfort. Wood and wood effects help to make the atmosphere even homelier.

095 90 20   030 50 60

添加重点色彩是设计师工作的一部分，为的是使氛围更和谐怡人。建筑方案所起的作用正如那些将光线引入舒适的、常用的双人床的圆形舷窗。在这里，圆形舷窗宛如有着不同视角的眼睛，从室内探向外界，为室内设计注入感性而异常舒适的元素。由于房间使用了红色和无烟煤色等独特的色调，形成许多对比，不断打破第一印象，显得格外真实。

光线和暖色调的、仿佛被阳光照亮一般明亮的墙壁融为一体。大多数人认为屋顶的斜坡并不让人生畏，因为它们的倾斜度和逐渐上升的高度能带来安全感和舒适感。木材的使用有助于营造更家常的氛围。

040 60 60   095 90 20

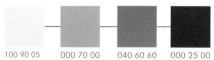

100 90 05   000 70 00   040 60 60   000 25 00

100 90 05   040 60 30   095 90 20

100 90 05   000 70 00   030 50 60   040 60 60

/ Chapter 5

ICONIC

标志意义

A chart as if from an innocently crazy flatshare, whose grandparents considered Woodstock worthy of imitation. Quite exciting, but possible.

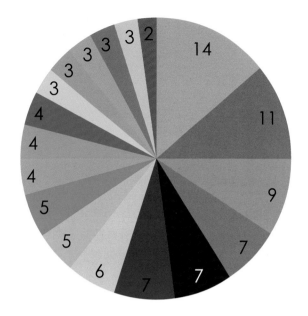

分格图使人想起那个天真疯狂的年代，祖辈们热衷于效仿伍德斯托克。令人异常兴奋的色彩风格，同时也蕴含了无限可能。

A colour stripe like a candy wrapper, or like the candy itself. It tastes very sweet, a little too pungent, and it colours our tongues and lips.

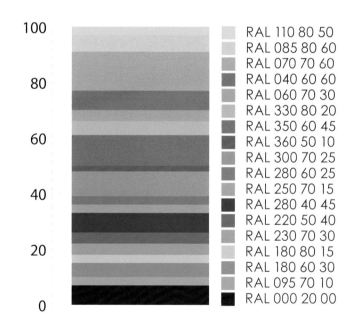

彩色条纹像糖纸，或像糖果本身。它甘甜却又有点齁人，而且有着与舌头和嘴唇一样的颜色。

Real movement provokes the iconic. Mainly lines and amorphous surface images.

真实的运动激发了图标的生成。主要为线性，表面图像并不规则。

ICONIC 标志意义

The relationship to culture is coincidental because the iconic is unintended; but it is play, amusement, coincidence, outlandishness, and has a great provocative potential. The iconic needs little – but it is beautiful if it is accompanied by narrative colours.

图标与文化的关系纯属巧合，因为它往往是偶然形成的，来自于游戏、娱乐活动、偶发事件和很多奇怪的潜在事物。图标并不需要什么意义，但如果赋予其叙事的色彩，它会格外美丽。

/ Chapter 5

FUTURISTIC

未来主义

The proportion of shades close to nature is barely visible. The idea of the future stimulates one's mind, which essentially does not want to get into emotional colour sequences.

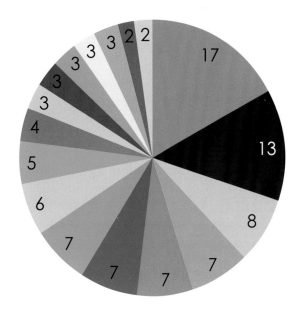

接近大自然的色调隐隐可见。关于未来的想法刺激着人们的头脑，但避免陷入情绪化的色彩序列。

The colouring of the bars is neither nice nor beautiful nor stimulating. The aspiration for objective clarification, which is garnished with a large pinch of scepticism, hovers above it.

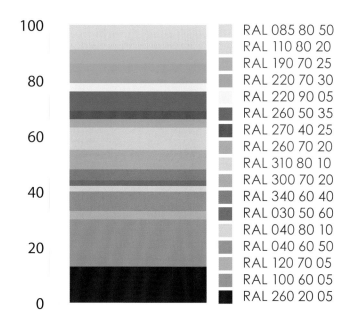

RAL 085 80 50
RAL 110 80 20
RAL 190 70 25
RAL 220 70 30
RAL 220 90 05
RAL 260 50 35
RAL 270 40 25
RAL 260 70 20
RAL 310 80 10
RAL 300 70 20
RAL 340 60 40
RAL 030 50 60
RAL 040 80 10
RAL 040 60 50
RAL 120 70 05
RAL 100 60 05
RAL 260 20 05

条形图的色彩既不漂亮也不美观，更不刺激。隐含着对客观特征的渴望，同时也存在怀疑。

A few graphic symbols can be recognised. The diagonals indicate a dynamic interpretation structure.

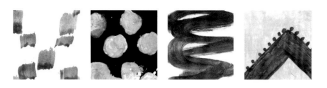

少数图形符号可以识别。对角线显示出动态的演绎结构。

FUTURISTIC　　未来主义

The colourfulness of the futuristic ranges from the fairly cool to the black and white. It is quite facile. The take-off from the present to the adventure of the future seems to incidentally trigger some anxious and dynamic colour and drawing processes.

未来主义风格的色彩包含了相当冷的色调和黑色、白色，这较易理解，脱离当下去未来冒险似乎在配色和绘画过程中都引发了一些焦虑。

/ Chapter 5

ECCENTRIC

离经叛道

The colour circle does not appear especially harmonious. The opposites hardly attract each other. The eccentric does not have to be loved, but can attract attention as an appealing alternative suggestion.

色盘看上去并不和谐。对立色无法互相吸引。离经叛道风格的色彩并不需要被人爱，但作为一个颇有吸引力的替代方案，它同样动人。

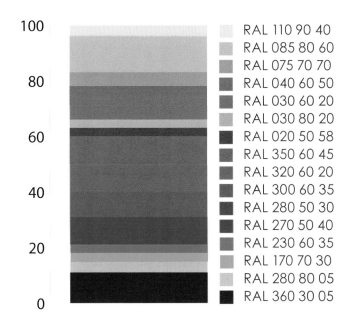

As soon as the colour sequence is straightened out, the enfant terrible turns into a fairly decent child. The dynamic content of the sequence still remains sufficiently visible.

RAL 110 90 40
RAL 085 80 60
RAL 075 70 70
RAL 040 60 50
RAL 030 60 20
RAL 030 80 20
RAL 020 50 58
RAL 350 60 45
RAL 320 60 20
RAL 300 60 35
RAL 280 50 30
RAL 270 50 40
RAL 230 60 35
RAL 170 70 30
RAL 280 80 05
RAL 360 30 05

只要把色彩顺序理顺，这个"顽童"看起来就会顺眼许多。此序列的动态变化清晰可见。

The test participants were encouraged to create excessive drawings. The large number of peaked, wave-shaped and zigzag lines are the result of the creative-sensual analysis.

试验参与者被鼓励尽量多地绘图。经过创意感性分析，得出了大量尖的、波浪形的、锯齿状的线条。

ECCENTRIC　　　　　　　　　　　离经叛道

With wild enthusiasm, the test participants immersed themselves in eccentricity. The most intense colours, piercing shapes and exalted zigzag or lightning patterns are used for celebrations.

试验者参与带着狂野的热情，不禁沉浸在这个离经叛道的世界里。最热烈的色彩、尖锐的形状和激昂的之字形或闪电图案可用于庆祝活动。

PROGRESSIVE

激进革新

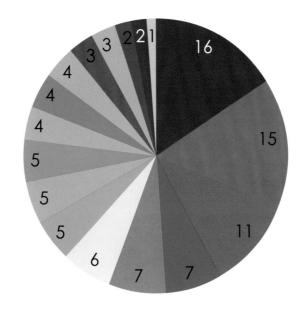

The large number of red and crisp blue nuances illustrates that dynamic mechanisms spur the progressive. In our opinion, they demonstrate the speed or acceleration of the process rather than the contents themselves.

大量的红色和清新的蓝色说明了动态机制推动革新。在我们看来，它们展现的是速度和加速的过程，而不是内容本身。

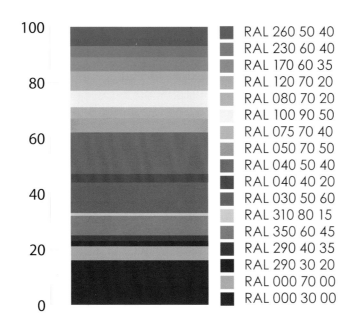

The colour sequence looks vibrant and stable. The test participants hardly attempted to represent anything searching or coincidental or creative.

色序看起来充满活力且稳定。试验参与者几乎没有任何探寻性的、随机的或有创意的尝试。

The symbols of this term actually look like immovable statements. Seriousness and a good deal of technical expertise cling to the progressive representations.

这个词的符号实际上看起来很稳定。激进革新风格的诠释与严肃的态度和大量的专业技术密切相联。

PROGRESSIVE　　　　激进革新

Progress does arrive quietly. As shown by the considerable share of black, we perceive future colours to be highly technical and not the result of creative-intellectual analysis.

革新确实悄然到来。另外，由于黑色占了相当大的比例，我们认为未来的色彩看起来极富技术感，而并非创造性智力分析的结果。

/ Chapter 5

ICONIC
标志意义

FUTURISTIC
未来主义

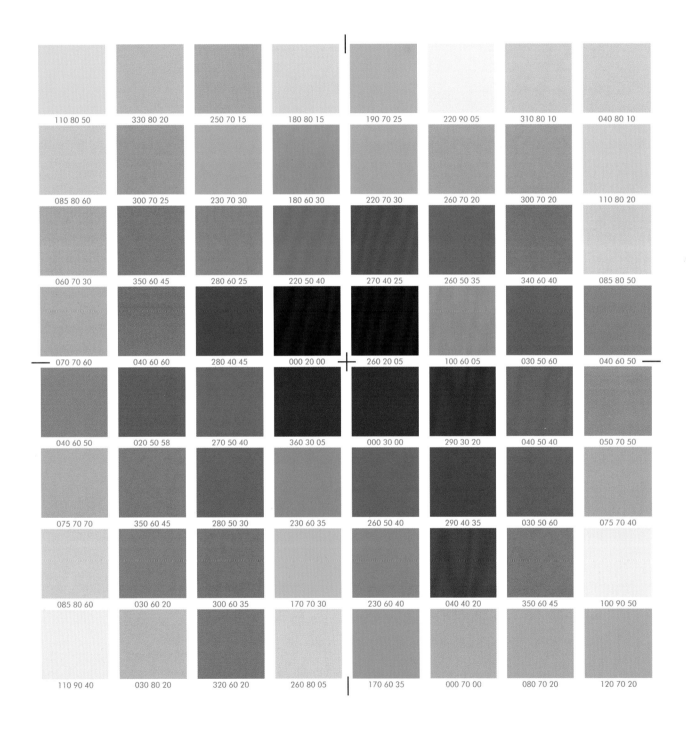

ECCENTRIC
离经叛道

PROGRESSIVE
激进革新

ICONIC
FUTURISTIC
ECCENTRIC
PROGRESSIVE

标志意义
未来主义
离经叛道
激进革新

The colours and signals are emphatic, accentuated and dynamic. As an overall theme, they sometimes strike us as too loud or too contradictory. However, what we expect of the future promotes the speculative. That which is appealing always lies ahead. We know the present well enough. Engagement with the future is essential because it prompts action and not reaction.

The shades are equipped with great luminosity, and the nature of the drawings is graphically linear and large in scale. The terms challenge one to make a statement, and they take a stand. Living in such a colourful environment will empower and sweeten the stay with power.

这些色彩和符号看起来热烈、突出且动感十足。在一个主题下，它们有时太花哨，有时反差太大，令我们吃惊。关于现在，我们知道的已经够多了。然而，我们对未来的预计充满了投机性，我们看到的总是有吸引力的东西。但联系未来非常有必要，因为这激起的是主动行动而不是被动反应。

这些色调与高亮度相搭配，这些图本质上多呈线性，规模也较大。这些词所描绘的色彩有自身的立场，也要求人们做出表态。在这样的色彩环境中生活，热烈而甜蜜。

FUTURISTIC 未来主义

Futuristic hotel design presents itself as bolide-like and dynamic. Arches and curves are the principal suppliers of the new design vocabulary. It has long outgrown the experimental stage. We have become accustomed to the soft shapes because we like the bionic approach. This could be the new functionalism with a formal style that is more sympathetic because it is softer. Or perhaps it is a new design vocabulary, which has modelled itself on human and faunistic structures and units. Muscles, sinews, body parts and body segments provide key words for a design philosophy that focuses on that which is functional for humans, and that which is very ergonomically formed. Its function-oriented features focus on individual requirements rather than on the straightforward optimisation of systemic effectiveness. The often neglected switches have an increasingly sensuous, haptic look and feel, which beckons with both haptic stimuli and technically enriched functions.

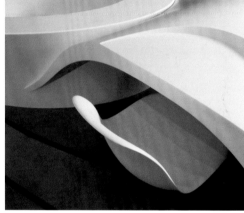

New materials make it possible to define the world of interiors in a different way than in the past. The big bathtub with a divan extension, which might – or should – be equipped with healthcare or wellness functions, is excellent. Perhaps it is an enormous fitness bed with an additional bathing function, and a work, culture and entertainment section integrated into the foot end.

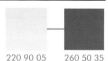

220 90 05 — 260 50 35

未来主义的酒店设计像火流星般充满活力和动感。未来主义早已超越试验阶段。拱形结构和曲线是这个设计新词的主要形式。我们习惯于柔和的形状，因为它们采用了仿生的方法。而这可能是一种全新而正式的功能主义，更柔和，更能引起共鸣。或者，这是一种新的设计语汇，仿照人类和动物本身的结构和单位。以肌肉、肌腱、身体部位和体节，为关注对人类来说实用且最符合人体工程学的设计理念提供关键词。它以功能为导向的特征使得它关注个性化需求而不仅仅是系统收益最优化。往往被忽视的开关设计越来越感性，它们的触感和外观昭示着它们既会带来触觉刺激且拥有更丰富的功能。

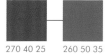
270 40 25   260 50 35

220 90 05   220 70 30   100 60 05   260 50 35

At first glance, the reception counter tells guests what to expect. It certainly looks exciting. Cosiness will not necessarily be the focus. The guests are in for something that is in tune with the times, cosmopolitan attitudes, progressivity, and, of course, futuristic flair. Every now and then it is wonderful to get into a time machine in order to get a whiff of the future.

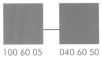

100 60 05　040 60 50

新材料使得以全新的方式定义室内世界成为可能。如带有长沙发椅的大浴缸已是极好了，但还有可能配备保健或健身功能。另一个非常棒的可能，大健身床增设额外的沐浴功能且集成其他，在足端附近便可进行工作、文娱等活动。

乍一看，接待台便告诉了客人接下来会发生什么。它看起来绝对令人振奋。设计焦点不一定在于提供舒适。客人们都乐于顺应时代潮流、国际化、技术进步，当然，以及未来主义的风格。不时进入时光机器、获得未来的气息会成为美妙的经历。

030 50 60

085 80 50

120 70 05

260 50 35

260 20 05

040 60 50　030 50 60

/ Chapter 5

WITTY

妙趣横生

The colours are snappy, quite funny, vivacious, and pure. Black serves only as a contrast and not as a surface area.

色彩明快，相当有趣、活泼而纯净。黑色只是作为对比色出现，不会占据很大比例。

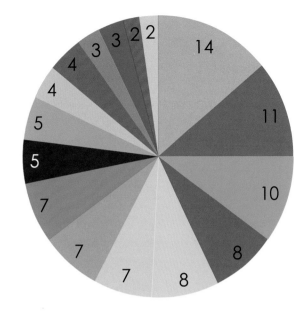

The bar chart comprises the hues of the rainbow. They come from the pastel area and so they have a pleasant luminosity.

条形图包含了五彩缤纷的色彩。它们色彩柔和，且拥有令人愉快的亮度。

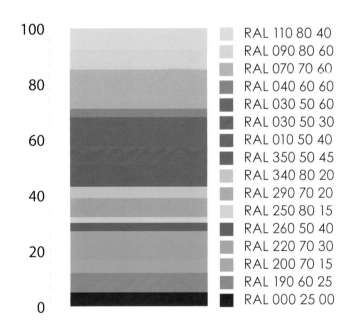

RAL 110 80 40
RAL 090 80 60
RAL 070 70 60
RAL 040 60 60
RAL 030 50 60
RAL 030 50 30
RAL 010 50 40
RAL 350 50 45
RAL 340 80 20
RAL 290 70 20
RAL 250 80 15
RAL 260 50 40
RAL 220 70 30
RAL 200 70 15
RAL 190 60 25
RAL 000 25 00

The term inspires graphic designs. About one half of the drawings is graphic, the other half is floral and dot oriented.

这个词能赋予图形设计以灵感，大约一半的图是规矩图形，而另一半则是花卉图案和波点。

WITTY　　　　　　　　　　　　　　　妙趣横生

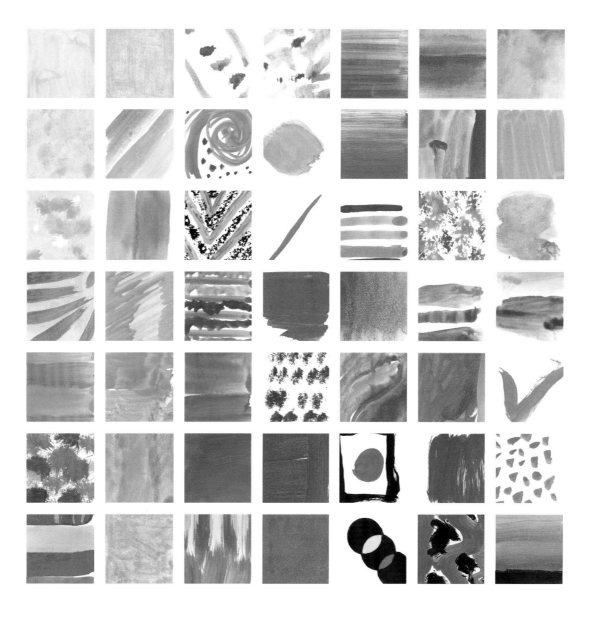

An apt description of the "witty": it is related to the clever, the sneaky, the shrewd and the savvy, and has considerable and undeniable charm. Something that looks witty enthuses us with its cunning in the form of fine, casual variety.

图示是对"妙趣横生"的贴切描述，妙趣横生往往和机灵、微妙、精明和富有洞察力相联，不可否认的是它极具魅力。看起来妙趣横生的物什富有灵气、形式多样而随意，令我们相当振奋。

/ Chapter 5

CHARMING

妩媚多姿

The colours look open, even transparent. About two thirds are warming but bright in tone. Almost all cool shades range from light blue or blue through to violet.

这些色彩有一种开放性，甚至看起来是透明的。大约三分之二的色调明亮而洋溢着暖意。几乎所有的浅蓝色或蓝色等冷色调都在向紫罗兰色转变。

Such colour sequences have reappeared in fashion. Fancy, strong South European apparel brands have, for example, revived these charming shades in their fashion and cosmetics collections.

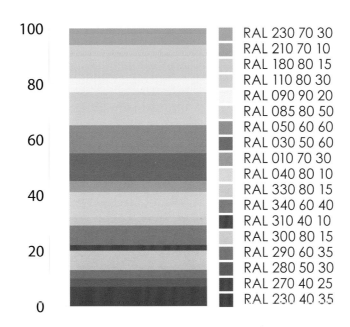

RAL 230 70 30
RAL 210 70 10
RAL 180 80 15
RAL 110 80 30
RAL 090 90 20
RAL 085 80 50
RAL 050 60 60
RAL 030 50 60
RAL 010 70 30
RAL 040 80 10
RAL 330 80 15
RAL 340 60 40
RAL 310 40 10
RAL 300 80 15
RAL 290 60 35
RAL 280 50 30
RAL 270 40 25
RAL 230 40 35

时尚界反复出现这种色彩序列。例如，一些花哨和南欧的著名品牌在它们的时装和化妆品系列中重新使用了这些迷人的色调。

A few diagonals, banana curves, circles, arches, parallel lines, and dots complement the drawing potential.

些许对角线、香蕉形曲线、圆圈、拱形结构、平行线和波点构成的图诠释了"妩媚多姿"风格的特点。

CHARMING   妩媚多姿

This colour scale radiates an aggressive, congenial, smiling openness. The hues are accompanied by matching flesh tones. They communicate typical human attitudes perceived as pleasant.

此风格色彩略显咄咄逼人，但它有一种开放性，仍然给人带来舒服的感觉。这些色调搭以深红色，可传达出愉悦的心情。

/ Chapter 5

PLAIN

平淡原味

Simple whitewashed nuances, larger proportions of grey and a bit of light blue, and simple green and dulled wooden shades suffice to give the term the corresponding colour atmosphere.

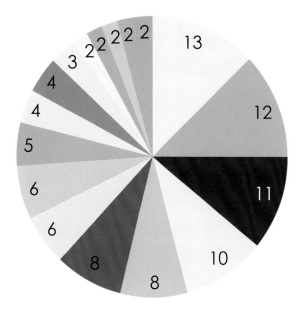

简单粉饰、深浅不一的色彩，更大比例的灰色和一点淡蓝色，还有简单的绿色和沉闷的木色调，足以描述这个词对应的色彩氛围。

The range has very little meaning, suiting a simple mind, a simple intellect or a simple idea. Mostly, plainness is not a strength but points out to a lack of dynamism and content.

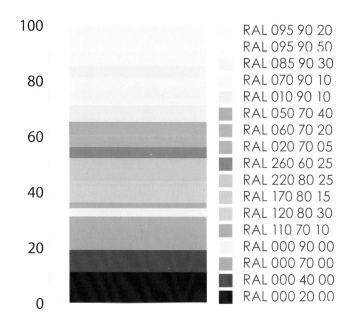

此风格色彩意义不大，只适合简单的情况、简单的头脑或简单的想法。大多数情况下，平实意味着缺乏力量，且缺乏活力和内涵。

The formal contents are graphic: a simple line or a plain-coloured, smooth surface are sufficient.

这种风格指向的是简单的线条、素色，还有光滑的表面。

PLAIN   平淡原味

Shining or aggressive are not part of the plain repertoire. Besides simplicity and an eschewal of emotional contents, the context of this term also includes attention to the puristic, material-oriented and unadorned.

闪耀和咄咄逼人并非平实原味的色彩给人的感觉。它除了看上去简单，以及有时给人隐藏情感的感觉外，它身处的环境往往表达了对纯粹、材质或了无修饰的关注。

/ Chapter 5

DISCREET

低调慎重

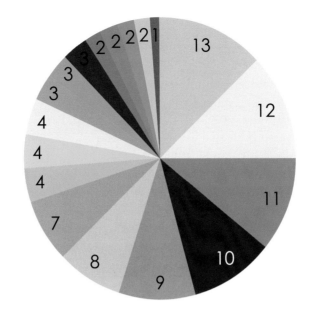

The corridors in the administration buildings of ailing companies, the staircases in nondescript 2-star hotels or in uncomfortable police barracks do not look too discreet, but they still look something like that.

境况不佳的公司的行政大楼的走廊，以及不伦不类的二星级酒店或不舒服的警察营房的楼梯，这些地方的色彩也许不能称为慎重，但确实符合这种不显眼的色彩概念。

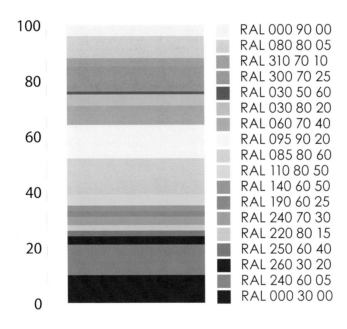

RAL 000 90 00
RAL 080 80 05
RAL 310 70 10
RAL 300 70 25
RAL 030 50 60
RAL 030 80 20
RAL 060 70 40
RAL 095 90 20
RAL 085 80 60
RAL 110 80 50
RAL 140 60 50
RAL 190 60 25
RAL 240 70 30
RAL 220 80 15
RAL 250 60 40
RAL 260 30 20
RAL 240 60 05
RAL 000 30 00

The colour sequence maintains its worn-out discreet charm which the beholder can only bear when in a good mood. One can see: in this overall aesthetic environment even the grey nuances appear enjoyable.

旁观者只有心情好的时候，才能忍受这种破旧而不显眼的色彩序列。人们可以看到，在这种整体的美学环境中，深浅不一的灰色看起来更舒服。

The discreet gets by with little formal content. An exuberance of graphic signs of emotion is rarely visible.

没有图形可言，几乎看不到任何体现情感的图形标志，低调色彩风格的图只是勉强可以接受。

DISCREET   低调慎重

The colours come in soft-footedly: Do not attract too much attention! That is why there is a large number of emaciated, dull, cool or featureless nuances. Black and grey serve as metaphors for going-into-hiding and invisibility.

这些色彩悄声走来，不会引起太多的关注！这就是为什么毫无神采、毫无光泽、冷冰冰或毫无特点的色彩大量存在。黑色与灰色颇能带来隐形和隐藏起来的感觉。

/ Chapter 5

WITTY
妙趣横生

CHARMING
妩媚多姿

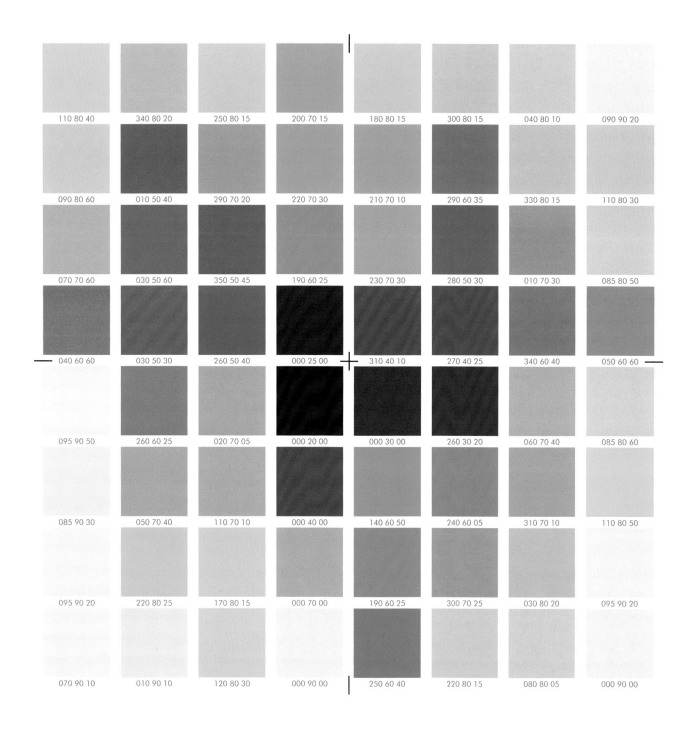

PLAIN
平淡原味

DISCREET
低调慎重

WITTY
CHARMING
PLAIN
DISCREET

妙趣横生
妩媚多姿
平淡原味
低调慎重

The "plain" colour range approaches the "discreet" one, whereby the former has a more non-committal nature than the latter. The "smart" and the "charming" ranges, on the other hand, radiate more medium-toned power. Combining the discreet with charm will help to achieve an elegant ambience. Combining the smart with the plain will lead to a few interesting contrasts.

Of course the thematic implementation of a single term is sufficient as a design idea. The plain should not eschew colours. It has, as we can see, stimulation potential in the range of calm limestone hues and slightly deeper medium tones. Seven dark accentuating shades give the entire colouring stability.

"平淡原味"风格色彩与"低调慎重"色系接近，与后者相比，前者更不明朗。另一方面，"灵气四溢"和"妩媚多姿"风格色彩，散发着更多中性的魅力。将低调慎重和妩媚多姿结合起来，有助于营造优雅的氛围。而灵气四溢和平淡原味的结合能带来一些有趣的对比。

当然，对于单个设计理念来说，实施一种色彩方案已足够。平淡原味的风格不应回避色彩。我们可以看到，它冷静的石灰石色调和略深的中性色调具有刺激作用。7种加重的暗色色调为整个着色带来稳定感。

CHARMING

妩媚多姿

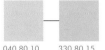
040 80 10   330 80 15

Pink, rosé, an elegant grey, and white that appears to be inspired by porcelain and alabaster. One comes across such rooms all too rarely, and yet they do exist beyond the city limits of Paris, Vienna and London. If we refer to something as "charming", it comes across with a smile and a wink. At its heart is a compliment, which comes from the heart.

The charming looks as genial as intended: a little playful, more feminine than masculine, light rather than heavy, and simultaneously sensitive, friendly and amiable.

Its appeal tends towards the courting of favour, high spirits, and elegance and lightness. Black and brown are as inappropriate as boulders, backslapping and shouting. Tea rose, perfume bottles, butterflies (the French term is better: papillons), and pink and rosé belong to the unmistakable characteristics of what we refer to as "charming".

粉红色、玫瑰色、优雅的灰色以及白色似乎受瓷器和雪花石膏的启发而来。这样的房间太罕见，但它们确实在巴黎、维也纳和伦敦之外也存在。如果我们用"妩媚多姿"来形容某物，想到的是微笑和眨眼，其核心是从内心深处发出的赞叹。

它跟预期的一样亲切，带点儿小俏皮，不那么阳刚，更柔和，明亮而不厚重，与此同时，给人细腻、友好、和蔼可亲的感觉。

它有着友好、积极向上、优雅和轻盈的魅力。黑色和白色、巨石、拍背大声鼓励和呐喊在这儿都不合适。香水月季、香水瓶、蝴蝶（法语词更贴切：PAPILLONS）、粉红色、玫瑰色毫无疑问都属于我们称之为"妩媚多姿"的风格。

090 90 20   040 80 10

090 90 20   300 80 15

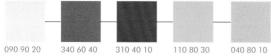

/ Chapter 6

HOTELS OF THE REGIONS: MEDITERRANEAN TO EAST EUROPEAN

This heading could in itself prompt the writing of dozens of new and interesting books on hotels. Wanderlust is, after all, just as strong as homesickness. At the very the edge of and also within Europe, there are hotels with an Asian, African or Latin American character, borrowing folkloristic elements from the many and varied regions of the world, or even embodying their distant origins in every detail. This goes beyond the authentic interior, extending to other services such as the specialities prepared in the kitchen and the offerings of the wellness, sports and spa areas.

As we become more and more familiar with international cuisines, we are also interested in the habits and customs of other peoples. We are prepared to adapt Italian, French, Spanish, and many variants of East Asian and Central and North American, recipes. Pizza's conquest of the world is just as convincing as that of French nouvelle cuisine or the triumph of the American Big Mac, the Turkish kebab and the intercontinental popularity of German and French sauerkraut. Fear of the unknown has long given way to curiosity about the unknown.

The exchange of cultures in art, architecture, design and science has become a prospering worldwide feature known as globalisation. It takes place not only as a non-material value or economic competition, but also as a platform for encounters that bring people together. As a matter of fact, hotels are the most important places for encounters as they are the interactive venues that promote contact. That is why the humaneness of their interior furnishing qualities is important. Cold splendour and technical functionality or unpleasant kitsch will never bring people together. We have come to realise that only a philanthropic design model, which focuses on human beings and their wishes, passions, hopes and individual needs, can work in the future. We simply have to want it to work.

地域风情酒店：
从地中海风格到斑斓东欧风

标题本身便能引发许多话题，足以写数十本关于酒店的书。毕竟，对旅行的酷爱如同乡愁一样强烈。欧洲边缘和欧洲内陆都有许多亚洲特色、非洲特色或拉美风的酒店，借用世界不同地区许多的民俗元素，甚至每一个细节都能体现它们遥远的起源。这超越了室内设计本身，扩展到其他服务，例如在厨房准备的菜肴、健身服务、运动区和水疗区。

随着我们越来越熟悉世界各地的美食，对其他民族的习俗也颇感兴趣。我们乐于尝试来自意大利、法国、西班牙和许多东亚、中美和北美的，但口味经本土改良的食物。比萨征服世界，法国新式菜肴的风行或美国巨无霸的胜利，土耳其烤肉和德国、法国泡菜的洲际普及都说明，对未知的恐惧早已让位给了好奇心。

伴随着全球化，文化艺术、建筑、设计和科学交流在世界范围内蓬勃发展。它不仅带来非物质价值或经济竞争，也为人们齐聚一堂提供了一个平台。实际上，酒店是最重要的邂逅场所，酒店里的互动有效促进了联系。这也是为什么有人情味的室内装修很重要。冷冰豪华、仅注重技术功能或媚俗的作品永远无法吸引人们齐聚一堂。我们已经意识到，唯有载满爱意的、关注人性以及人们期许、激情、希望和个性化需求的设计模式，才能适应未来。我们只想要能满足情感需求等的设计。

400	Mediterranean - Oriental - South American - African	地中海风 – 东方情调 – 奔放南美 – 粗犷非洲
416	North European - British - Central European - East European	纯美北欧 – 纯正英式 – 唯美中欧 – 斑斓东欧

/ Chapter 6

MEDITERRANEAN

地中海风

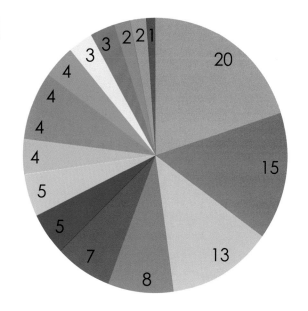

The warm colours of the landscape obtain their special appeal from the loud blue shades of the sky and the water. They are contrasted by the hues of the different greens of the flora.

象征着地中海的天空和大海、暖暖的蔚蓝色调有着特有的魅力。植物各种不一的绿色色调与其形成反差。

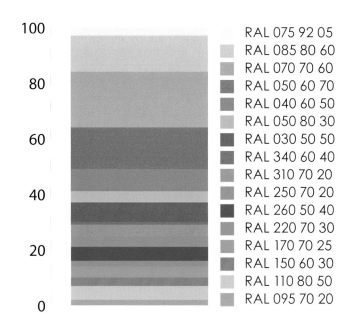

Only when travelling to the Mediterranean do the colours regain their clarity. The radiant sun reveals the colours' glory.

只有在地中海，色彩才会这么明艳照人，光芒四射的太阳揭示了色彩的荣耀。

RAL 075 92 05
RAL 085 80 60
RAL 070 70 60
RAL 050 60 70
RAL 040 60 50
RAL 050 80 30
RAL 030 50 50
RAL 340 60 40
RAL 310 70 20
RAL 250 70 20
RAL 260 50 40
RAL 220 70 30
RAL 170 70 25
RAL 150 60 30
RAL 110 80 50
RAL 095 70 20

Frequently, only a few vibrant areas suggest the Mediterranean, far from any graphic drama.

往往是图中那些活力十足的区域，而不是丰富而戏剧性的画面，使人想到地中海。

MEDITERRANEAN　　　地中海风

An atmosphere like on Majorca! The soft hues of the coarse-grained plasterwork and painted quarrystones consisting of irregular polygons, bright sand and medium-toned woods characterise the appearance of the western Mediterranean.

色彩鲜艳的沙粒、中性色调的木头、柔和色调的粗灰泥和不规则多边形的着色粗石，描绘出西地中海的特征。

ORIENTAL

东方情调

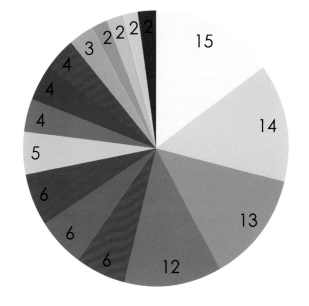

Almost exclusively sunny or precious colour shades characterise the range: from radiant gold to orange to purple and violet. Furthermore, there are precious lapis lazuli, cool malachite and chrome-green nuances.

这一色彩组合的特点在于囊括了非常独特而珍贵的颜色：从光芒四射的金色到橙色，再到紫色和紫罗兰色。不仅如此，还有十分珍贵的天青色、静谧的孔雀石绿以及铬绿色。

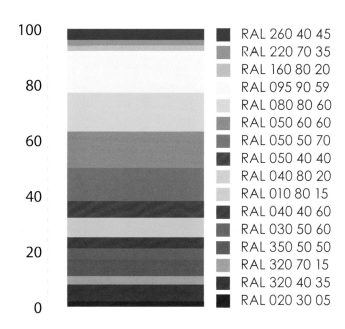

- RAL 260 40 45
- RAL 220 70 35
- RAL 160 80 20
- RAL 095 90 59
- RAL 080 80 60
- RAL 050 60 60
- RAL 050 50 70
- RAL 050 40 40
- RAL 040 80 20
- RAL 010 80 15
- RAL 040 40 60
- RAL 030 50 60
- RAL 350 50 50
- RAL 320 70 15
- RAL 320 40 35
- RAL 020 30 05

Colours as we know them from traditional Ottoman silk embroideries, which adorned festive and state robes. The motifs consist of tendrils and blossoms, which form an ornamental embellishment.

我们从传统的土耳其丝绸刺绣中获得这些颜色，它们被用在装饰节日庆典和制作国家礼服上。刺绣的图案包括卷须状或是描绘繁花盛开的景象，这些特点都使其极具装饰性。

The drawings are ornamental. They are reminiscent of classic textiles and ceramic tiles.

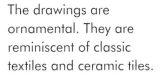

这些绘画充满装饰性。它们让人联想起古典织物和瓷砖。

ORIENTAL   东方情调

Pictures that awaken old memories of the Tales of 1001 Nights. Scheherazade, with gold interwoven silk fabrics flowing around her, dances to exotic sounds. Gold, lapis lazuli and emeralds blink on purple-coloured velvets, brocades and veils.

这些图勾起了我们心底对《一千零一夜》故事的回忆。我们仿佛看到谢赫拉莎德周身包覆金色交织的丝绸，随着充满异国情调的歌曲翩翩起舞。金色、天青色、翡翠绿在紫色丝绒、锦缎和面纱中时隐时现、熠熠生辉。

/ Chapter 6

SOUTH AMERICAN

奔放南美

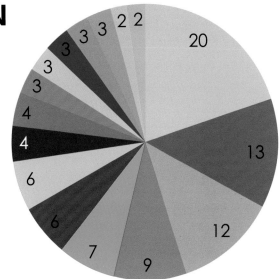

This is a picture the Central Europeans have of South America. It is shaped by media of all sorts, by hearsay and personal experience.

这反映出中欧地区的人们对南美洲的色彩印象。这种概念的形成来自多种媒介，或者是道听途说，或者是亲身经历。

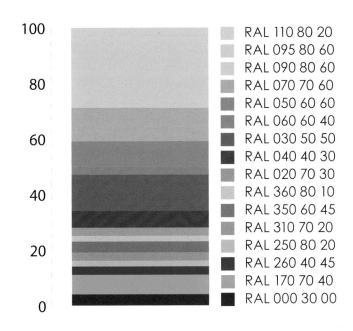

RAL 110 80 20
RAL 095 80 60
RAL 090 80 60
RAL 070 70 60
RAL 050 60 60
RAL 060 60 40
RAL 030 50 50
RAL 040 40 30
RAL 020 70 30
RAL 360 80 10
RAL 350 60 45
RAL 310 70 20
RAL 250 80 20
RAL 260 40 45
RAL 170 70 40
RAL 000 30 00

A colour range made by the sun. Even the turquoise and water blue seem to be more animated by rays of the sun than they could ever be in cooler regions.

这是属于阳光的色段。即便是蓝绿色和水蓝色，较之在较冷地区的色调，这里因为有了阳光的照耀，显得更加生机勃勃。

The paintings by the test participants are reminiscent of textile patterns: free from floral appliqués, but as colourful as T-shirts.

试验参与者画出的这些图案让人联想到了纺织品的纹理和图案，但却非素淡的缝花布料，而是色彩缤纷的T恤。

SOUTH AMERICAN 奔放南美

The Hispano-Portuguese-influenced southern part of America wears the colours of warm-hearted joie de vivre. Shades that were influenced by people and their enthusiasm for nature, the rivers and mountains, the sea and tropical forests and desert landscapes.

在西班牙语和葡萄牙语影响下的美洲南部为自己批上了一层温暖且趣味横生的彩衣。这些色彩是对这个区域的人们和他们对山川河流、大海雨林、沙漠景观等自然由衷热爱的最好体现。

AFRICAN

粗犷非洲

It is predominantly desert, clay and earth colours that cross people's minds when thinking of the African continent. Natural and folk art products also determine the colours.

非洲大陆带给人们的色彩印象主要为沙漠、黏土和土壤的颜色。一些自然的或者民俗艺术品也同样参与形成了人们对其的色彩印象。

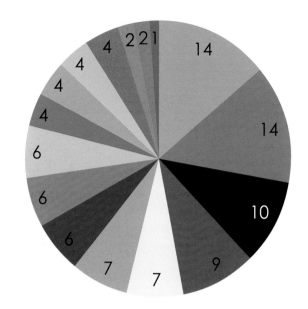

The shades have a lot to do with cosiness. They convey feelings of warmth and feelings of immediate proximity and friendship.

这些色调带给人们无比的舒适感。它们给人以温暖、亲近和友好之感。

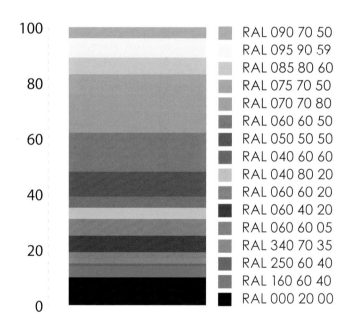

RAL 090 70 50
RAL 095 90 59
RAL 085 80 60
RAL 075 70 50
RAL 070 70 80
RAL 060 60 50
RAL 050 50 50
RAL 040 60 60
RAL 040 80 20
RAL 060 60 20
RAL 060 40 20
RAL 060 60 05
RAL 340 70 35
RAL 250 60 40
RAL 160 60 40
RAL 000 20 00

Some of the test participants have incorporated typically African symbols. These include serpentine lines, zigzags, meandering patterns, and colourful dot and dab motifs.

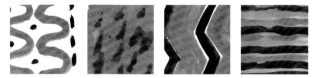

显然，一些试验参与者脑中已经有了非洲的典型符号意象。这些意象包括蛇形线条、锯齿状折线、形态蜿蜒的图案以及色彩斑斓的点状或随性涂抹的纹路。

AFRICAN

粗犷非洲

That's what Africa looks like. Exactly as it is presented in the brochures about regions from the Serengeti to the Sahara, from South Africa to Zambia. Lots of reddish earth, yellow sludge and ochre-coloured steppe with a few aqua-coloured brighter areas.

这就是非洲的样子！它与宣传画册中描绘塞伦盖蒂到撒哈拉沙漠、南非到赞比亚的图片几乎吻合。大片泛红的土地、黄色的泥土、赭石色的干草原之间又间或出现点点泛光的水蓝色。

/ Chapter 6

MEDITERRANEAN
地中海风

ORIENTAL
东方情调

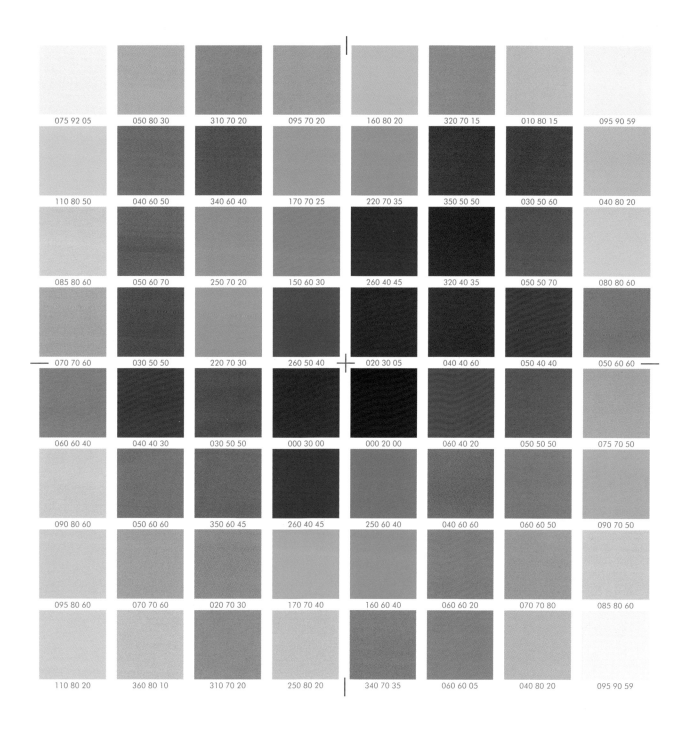

SOUTH AMERICAN
奔放南美

AFRICAN
粗犷非洲

MEDITERRANEAN
ORIENTAL
SOUTH AMERICAN
AFRICAN

地中海风
东方情调
奔放南美
粗犷非洲

The colours of the regions have their own authentic characteristics. Our ideas of the "foreign colour spectrum" are relatively realistic. The images of folkloristic architecture, design positions and historic features are represented in a nuanced manner. Participants in the study achieve a remarkable degree of agreement regarding their coloured points of view.

At least since the fifteenth and sixteenth century, the colour and formal contents of foreign cultures have influenced European cultures. In the fifteenth century, this began with South American motifs, followed by Turkish-Arabian motifs in the sixteenth century and, in the seventeenth and eighteenth century, Chinese and Indian, Middle Eastern and Japanese motifs. Egyptian elements were added at the dawn of the nineteenth century, followed by African and Mediterranean influences.

这些区域的色彩有着它们各自的特性。我们对于"外来色谱"的想法相对实际。民间建筑、设计定位和历史特色都能以色彩的方式一一呈现。研究参与者们对这些"有色"的观点达成了很大程度的统一。

至少在15和16世纪，外国文化的色彩和内容确实影响着欧洲文化。这种影响以15世纪的南美洲纹案为起始，紧接着是16世纪的土耳其-阿拉伯纹案，17和18世纪的中国和印度、中东和日本纹案。埃及元素在19世纪末逐渐融入，紧随其后又受到非洲和地中海文化的影响。

ORIENTAL   东方情调

050 40 40   320 40 35   350 50 50

350 50 50   040 80 20   320 40 35

Whether we are talking about hotels in the Asian, African or Moroccan style, we choose them because we are fascinated by their folkloristic momentum. We love the warm colours of the wood and the walls, and the rooms finished with rugs or ornamental tiles or coloured marble. The bath in an oriental hammam is also more beautiful than any European bathtub. In the same way, the outdoor bathroom of a comfortable Chinese hotel refreshes us more than an austere shower.

We enjoy sitting in rocking chairs in the foyer or lobby or in the living room of a suite whilst chatting and sipping a local drink or the usual Scotch. Just add the appropriate folk music and the guests will feel at ease.

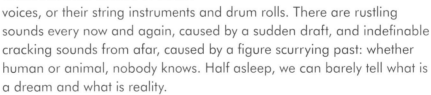

The light from the perforated brass lampshades, which are suspended from the ceiling or mounted on the wall, conjure a filtered glimmer of light radiating from many diffuse sources. Until late into the night, when the temperature becomes bearable, we enjoy the noises of the night that never cease. They are continuous: the babble of human voices, or their string instruments and drum rolls. There are rustling sounds every now and again, caused by a sudden draft, and indefinable cracking sounds from afar, caused by a figure scurrying past: whether human or animal, nobody knows. Half asleep, we can barely tell what is a dream and what is reality.

We actually love the transitions from one culture to another, from stylistically reassuring unity to foreign additions, displaying the results of one's personal collecting passion and the many successful attempts to bring together East and West, South and North, different kinds of folklore, history and a variety of stories.

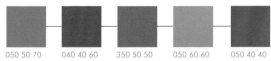

050 50 70　040 40 60　350 50 50　050 60 60　050 40 40

无论我们是谈及亚洲、非洲或是摩洛哥式的酒店，选择它们是因为我们对其民族魅力的神往。我们热爱木材和墙面的暖色，热爱铺设着地毯、装饰瓷砖和彩色大理石的地面。在东方土耳其风格的澡堂沐浴常常比躺在欧式浴缸中更惬意。同样，在一家舒适的中国酒店的户外澡堂中沐浴则比简单的淋浴更让人神清气爽。

我们享受坐在前厅或大堂石椅上休息时的自在，享受在套房的起居室里一边畅聊、一边喝着当地饮料或是苏格兰威士忌时的悠闲。若能再增添一点点的民谣，则美妙升级，氛围也愈加轻松。

光线经悬挂于天花或是固定在墙面的带孔黄铜灯罩滤出，各处光源散射的光线营造出微光闪烁的效果。夜深时，当气温变得不那么难耐，我们享受起永不止息的属于夜晚的"嘈杂"：人们的喃喃声、弦乐或是圆鼓奏起的声音。一阵疾风吹过，或是远方难以分辨的破裂声，或是不知是人还是动物疾走而过的响声，这些沙沙声总时不时作响。酣睡至半，我们几乎无法分辨这是梦境还是现实。

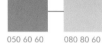

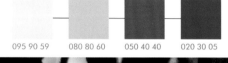

050 40 40   050 60 60

080 80 60

095 90 59

050 40 40

050 60 60

040 40 60

我们真的很喜欢文化与文化之间的过渡，从一种制式的、令人心安的本体到另外的文化附加物，展示出人们对广泛收集的热爱。人们在融合东西南北不同的文化、历史和多彩的故事方面做出了一次次成功的尝试。

AFRICAN

粗犷非洲

060 60 50   060 40 20

Our rooms have stable shutters and doors. Everything gives the impression of solidity; even the high beds with totems and charms on the headboard protect us. Every now and again, we come across colonial-style ambiences, which are enriched with authentic furniture and surface elements, genuine furnishings and precious exotic furniture and cultural objects dating from the last two centuries.

The impression for "African" aims at more than just an authentic ambience. It is about invoking an exotic atmosphere: elephant trunks, lion heads as towel holders and drum designs for the washbasin. The vertical rods are martial ritual symbols. We have always loved and will continue to enjoy integrating the mysterious other into our own symbolic worlds. In this way we can doze and dream blissfully in the "African" bed guarded by spears.

我们的房间有着牢固的百叶窗和门。室内的每一件物品都透露出一种坚固之感；甚至那高高的床，床头饰有图腾和符咒以保护屋主，同样传达着坚固的概念。我们不时遇到一些殖民地时期风格的场景，内部布置耐用的家具和一些平面装饰元素，真材实料的陈设物品和珍贵的异国物件，以及长达百年历史的文物。

非洲印象不只是一种非常真实的场景，它还能唤起一种异国情调：象牙、狮头形态的布巾架子以及以非洲鼓为设计造型的浴缸。垂直放置的木棒是一种尚武的象征。我们总是感到欢欣并且愿意继续享受这种将外来神秘意象融入我们自己的符号世界的美妙过程。如此，我们便能安稳地、愉悦地在符咒保护的非洲大床上小憩或是慢慢进入梦乡。

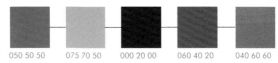

/ Chapter 6

NORTH EUROPEAN

纯美北欧

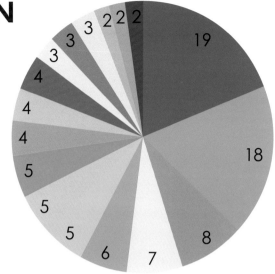

The varied nuances from sea blue through to fjord and pond-green colourings show the spectrum and creativity-releasing appearance of the colour blue. It is obvious: blue is anything but boring.

从深浅渐变的海蓝色到峡湾绿和池塘绿，这一色轮组合彰显着蓝色的创造性。显而易见，蓝色绝不会让人感到枯燥乏味。

The small number of playing colours also shows how much we agree with the statement that in Northern Europe, blue prevails, with a few green, vernal sprinkles.

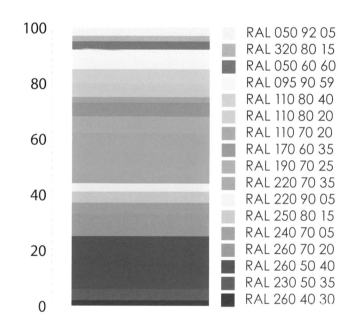

RAL 050 92 05
RAL 320 80 15
RAL 050 60 60
RAL 095 90 59
RAL 110 80 40
RAL 110 80 20
RAL 110 70 20
RAL 170 60 35
RAL 190 70 25
RAL 220 70 35
RAL 220 90 05
RAL 250 80 15
RAL 240 70 05
RAL 260 70 20
RAL 260 50 40
RAL 230 50 35
RAL 260 40 30

图中少量逗趣的色彩再次佐证了这一观点：在北欧，蓝色盛行，点缀着些许绿色，如春雨般洒落。

Light, swirling movements of the air, firm water colours aillumination by a clear spells of a cool winter sun.

这些图让人联想到轻盈飞旋着的空气，以及静静流淌的水，在清冷冬日里的阳光下熠熠闪光。

NORTH EUROPEAN 纯美北欧

No surprise: plenty of water blue, bright leaf and meadow green and wintery, foggy grey. Here in the north, we expect the colours of clear air, of ice, water and bone-chilling cold. The few dabs of colour are perhaps faded memories of Scandinavian timber cottages.

不用惊讶：这里有满眼的水蓝，明快的树叶和草地的绿色，以及冬季中雾似的灰色。在北欧，我们期待的色彩来自清透的空气、冰、水以及刺骨的寒冷。还有几抹色彩也许来自记忆深处模糊的斯堪的纳维亚式的小木屋。

/ Chapter 6

BRITISH

A colour image of great power. Designed with heart and reason in equal measure. Red heroism and blue calculation form a symbiotic unit.

纯正英式

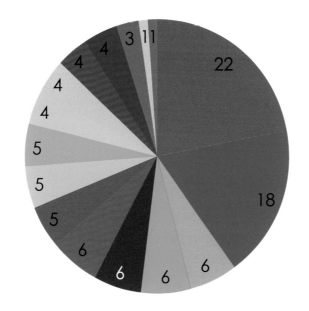

这张色彩图力量勃发，一半充满情感，一半彰显理性。英雄主义的红色和审慎的蓝色形成一种异类共生的组合。

An effective bar graph has been created here. It has a tangible acoustic value and an evocative power. A precious item in this sort of wrapping appears twice as valuable.

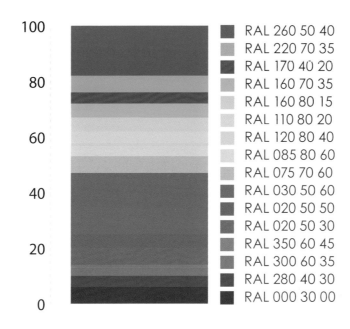

这是一张效果显著的条形图。它似乎可听可感，并具有唤起人回忆的力量。贵重的物品通过这一色系的包覆价值倍增。

The symbols are influenced by heraldry, too. The vertical and horizontal bars indicate stable persistence or dynamic striving.

这些符号的绘制也受到了纹章学的影响。垂直和水平条状显示出一种持久稳固或是积极进取的寓意。

BRITISH   纯正英式

In these images, Britain's colours have been victorious. Not only on celebration days public life is determined by red and blue; both shades are omnipresent on busses, phone boxes and underground logos.

在这些图片中，大不列颠的色彩奏响一支胜利之歌。在那些值得庆祝的节日中，红色和蓝色不只主导着公共生活，还频繁出现在公车、电话亭和地下通道标识中。

/ Chapter 6

CENTRAL EUROPEAN

The cooler shades form a majority in the colour spectrum. Dissonances are barely visible. It seems that all residents love nature, the water, and a very few clear spells when the sun is shining occasionally.

唯美中欧

冷色调主导着这一色彩组合，视觉上十分和谐。似乎这一地区所有的民众都热爱自然、水和偶尔露脸的灿烂阳光。

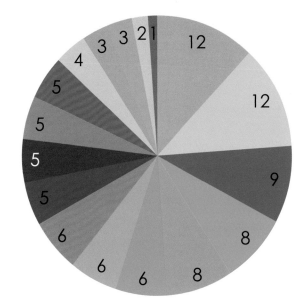

The colour range is reminiscent of park rather than factory landscapes. Whether the association is wishful thinking or reality is debatable.

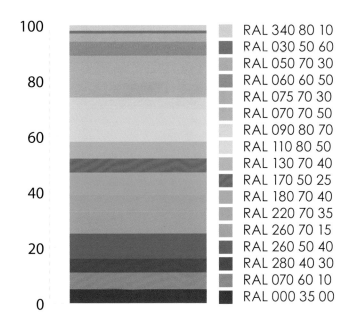

这个色彩组合能唤起人们对公园的印象，而非工厂周边的景观。不过，这种联想到底是一种一厢情愿的主观看法还是实际情况呢，有待商榷。

The statements on Europe's colours are more emphatic than those on formal contents. These display few characteristic signs or symbols.

欧洲色彩在一些非正式的题材中更能发挥功用。它们很少用来展示一些特色的标识和符号。

CENTRAL EUROPEAN 唯美中欧

It is hard to believe, but true, that Central Europe can make do without exhilarating and heart-warming red. The watery, herbal, sunny, and a few clayey nuances constitute the colour repertoire of the centre of the folkloristic and geographic characteristics of this small part of the world.

也许很难令人信服，但实际上中欧并不需要特别令人振奋或暖心的红色。这一片充满民俗气息和富于地域特色的小小区域，静躺在这偌大世界的一隅，水、草、阳光和些许泥土的色调就足以构成这片区域的全部色彩。

/ Chapter 6

EAST EUROPEAN

斑斓东欧

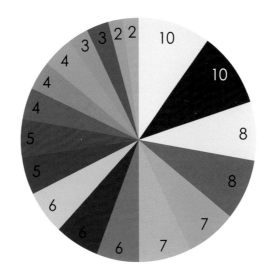

The few green shades originate from Russia's folklore; a few vibrant Turkic colours from the south-eastern Balkan are also present.

这一色盘中,少量的绿色调源自俄罗斯民俗;色盘中同时呈现的还有源自巴尔干半岛东南部那充满突厥民族风情的色彩。

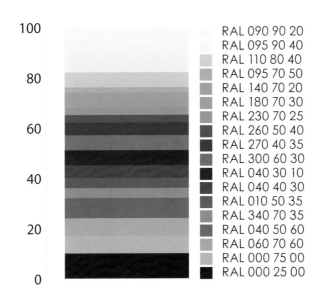

RAL 090 90 20
RAL 095 90 40
RAL 110 80 40
RAL 095 70 50
RAL 140 70 20
RAL 180 70 30
RAL 230 70 25
RAL 260 50 40
RAL 270 40 35
RAL 300 60 30
RAL 040 30 10
RAL 040 40 30
RAL 010 50 35
RAL 340 70 35
RAL 040 50 60
RAL 060 70 60
RAL 000 75 00
RAL 000 25 00

This colour sequence is quite loud. It is reminiscent of wonderful floral folk art embroideries from Poland, the Czech Republic and Slovakia.

这一色彩序列相当"喧闹"。它让人们想起那些美妙的民间花卉刺绣艺术,那些来自波兰、捷克和斯洛文尼亚的精湛手工艺。

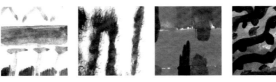

The drawings consist of graphic elements, stripes, bars, and grids.

这些绘图包含着诸如带状、柱状、网格等图形元素。

EAST EUROPEAN 斑斓东欧

Eastern and western impressions combine to form a picture of clear Baltic and warmer South European Balkan influences, together with a few historic Austrian, imperial yellow-golden colour types.

将东西欧印象组合而观，在这样一幅图上，我们看到了波罗的海的纯净，欧洲南部温暖的巴尔干半岛的影响，以及些许充满历史气息、展现奥地利威严的金黄色。

NORTH EUROPEAN
纯美北欧

BRITISH
纯正英式

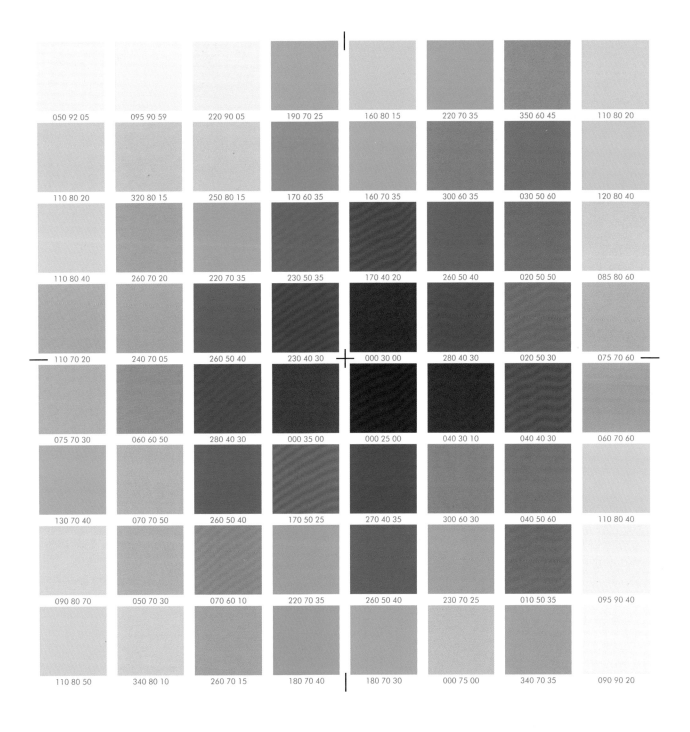

CENTRAL EUROPEAN
唯美中欧

EAST EUROPEAN
斑斓东欧

NORTH EUROPEAN
BRITISH
CENTRAL EUROPEAN
EAST EUROPEAN

纯美北欧
纯正英式
唯美中欧
斑斓东欧

A nuanced image of the colouristic character of a part of Europe becomes visible. One of the fundamental cultural strengths of the entire continent manifests in its heterogeneity, as proved by the geographic colour map. When one adds the Mediterranean countries, the picture becomes even more colourful.

The Central European part is relatively ponderous compared to colour-rich Britain. In our eyes, Eastern Europe also looks very colourful, accentuated with more mystical shades than there are in the western part. Blue and green-coloured shades form the main part of the European colour identity.

一张细致入微的色彩过渡图让一部分欧洲的典型色彩立刻映入眼帘。对于整个欧洲大陆而言，它的文化底蕴和精髓之一在于它的丰富多样。大陆的地理色彩分布图能够清晰明了地展示这种丰富。当人们把地中海国家也纳入这张地图中，色彩则变得愈加缤纷。

与色彩丰富的大不列颠相比，中欧似乎略显呆板。在我们眼中，东欧同样五彩缤纷，尤其它独有的神秘色彩较之西欧，更胜一筹。蓝绿色调成为欧洲主要区域的色彩标志。

NORTH EUROPEAN 纯美北欧

230 50 35   050 92 05   260 70 20

We entrust the cool water-and-sky and winter colours primarily to those from Northern Europe. For the maritime states with long coastlines and beaches in particular the colour blue is of great importance. However, we also know that vibrant red shades, green and brown nuances are widely used in combination with blue, especially in interiors, but also as architectural colours.

One way and another, our wishes and expectations are focused on blue. Especially on the cylindrical ovens still in use today, the tiles are mostly white with blue drawings and brass fittings. The blueprints of bygone eras are always combined with shades of white. Sometimes, white is the font colour, and at other times it is a classic navy blue. Wood used for the interior fittings and pieces of furniture are frequently painted white, too. In more elegant houses, bright poplar and birch wood have had a dominant position for generations as a veneer and in the form of solid wood. The pleasantly cool colourfulness of classic German spa-style architecture with a captain's room and panels painted white is enjoying an impressive renaissance, giving the impression of exclusivity, luxury and homeliness.

清爽的水天色和冰凉的冬季色彩似乎是北欧与生俱来的。那些有着长长海岸线和沙滩的沿海城镇尤其如此，在那里，蓝色扮演着极其重要的角色。亮红色调及绿色、棕色常常与蓝色结合使用，这种搭配常见于室内设计领域。当然，这也是一种普遍的建筑色彩方案。

总的来看，我们的期望还是要由蓝色实现。特别是柱式烤炉还在使用的今天，铺砖大部分为白底搭配蓝色图绘，烤炉配件则为黄铜色。那些逝去岁月总是与白色相伴。一些时代，白色是普遍的字体用色；一些时代则使用经典的海军蓝。那些被用作室内配饰及用于家具制作的木头也常常被刷上白色漆。德国经典SPA风格建筑有船长室和白色甲板，其令人愉悦的清爽用色给人留下尊贵、奢华和家常般亲切的深刻印象。

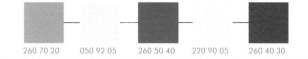

260 70 20   050 92 05   260 50 40   220 90 05   260 40 30

ACKNOWLEDGEMENTS AND EXPRESSION OF THANKS

We cordially thank the persons listed below for their colour interpretations, which are represented in this book in full colour on the white page. They created the scientific base for this project. Some of them are familiar with work in a hotel from a professional point of view; most, however, have been frequent or at least occasional guests in the microcosm of hospitality. Their opinions have been coloured by memories of visits to hotels or by professional experience. The names of these objective, creative and highly talented interpreters are:

Anne Alpers-Janke, Gerd Arp, Elke Becker, Claudia Biederbick, Stefanie Biermann, Larissa Billinger, Wiebke Biß, Lea Bohm, Heinrich Böhm, Wolfgang Bonin, Tobias Brown, Imme Chaaßen, Isgard Eikmeier, Fabian Engels, Alexander Fitz, Dietrich Floor, Regina Gehlich, Mareike Godeck, G. Godeck, Annette Hartmann, Inga Heder, Paula Hein, Sofia Hein, Carsten Hennig, Peter Joehnk, Felix Jöhnk, Stefan Jöhnk, Caroline Kick, Carlies Krause, Corinna Kretschmar-Joehnk, Anne Maaßen, Andreas Masel, Thomas Mittendorf, Thomas Notholt, Frank Oehmichen, Annika Petersen, Harry Petersen, Natalia Platanovitch, Else Roos, Maren Roos, Janina Rosky, Andy Rudolph, Birgit Schallock, Karl Schlichting, Anne Scholz, Thomas Scholz, Hilde Schultz, Sandra Schuster, Iris Schuth-Kniphals, Carola Simons, Dietrich Venn, Jana Vonofakos, Friederike Warrelmann, Kathleen Webersinke, Hans-Jürgen Werndt, Ursula Werndt, Cathrin Wißmann, Doris Zimmermann

The authors admire the enthusiasm of all silent and active helpers who have supported this project.

致谢

我们诚挚地感谢以下人士，他们为这本书的色彩研究做出了极大的贡献。这本书为全彩印刷，与色彩相关，是他们为这本书、这次研究奠定了科学基础。他们中有一些人非常熟悉酒店行业，视角非常专业；但大多数是经验丰富的酒店顾客或偶尔入住酒店的客人，作为酒店客人的一个缩影，他们提供了各种关于酒店的回忆或专业经验，着实富有创意和才华：

Anne Alpers-Janke, Gerd Arp, Elke Becker, Claudia Biederbick, Stefanie Biermann, Larissa Billinger, Wiebke Biß, Lea Bohm, Heinrich Böhm, Wolfgang Bonin, Tobias Brown, Imme Chaaßen, Isgard Eikmeier, Fabian Engels, Alexander Fitz, Dietrich Floor, Regina Gehlich, Mareike Godeck, G. Godeck, Annette Hartmann, Inga Heder, Paula Hein, Sofia Hein, Carsten Hennig, Peter Joehnk, Felix Jöhnk, Stefan Jöhnk, Caroline Kick, Carlies Krause, Corinna Kretschmar-Joehnk, Anne Maaßen, Andreas Masel, Thomas Mittendorf, Thomas Notholt, Frank Oehmichen, Annika Petersen, Harry Petersen, Natalia Platanovitch, Else Roos, Maren Roos, Janina Rosky, Andy Rudolph, Birgit Schallock, Karl Schlichting, Anne Scholz, Thomas Scholz, Hilde Schultz, Sandra Schuster, Iris Schuth-Kniphals, Carola Simons, Dietrich Venn, Jana Vonofakos, Friederike Warrelmann, Kathleen Webersinke, Hans-Jürgen Werndt, Ursula Werndt, Cathrin Wißmann, Doris Zimmermann.

对这些默默支持或积极热情的参与者们，深表感谢。

THE AUTHORS   作者

Axel Venn studied design and free composition at the Folkwang School of Design in Essen. He works for international companies as a creative and scientific consultant. Prof. Axel Venn lives and researches in Berlin. He has published twenty-five books on colour in science, marketing and trend style as well as in experimental disciplines. So far, his books have been translated into twelve languages.

Janina Venn-Rosky graduated in Design at the University of Fine Arts (UFBK) in Hamburg. As a co-author she has so far published seven books and fifteen complex trend books. For many years, she and her husband, Dipl. Sozialökonom Alexander Venn, have collaborated closely with Axel Venn, especially in the field of empirical research and its practice-related applications.

Corinna Kretschmar-Joehnk, co-managing director of JOI-Design GmbH, is one of the industry's foremost experts in hospitality interior architecture and design. She studied art history in Würzburg and interior design in Detmold. For over 20 years, her visionary concepts for both major hotel brands and independent boutiques have delighted guests and shaped trends around the globe. She and her husband, Peter Joehnk, have managed their leading European interior architecture studio JOI-Design since 2003. The duo has also shared their passion for creativity in two previously published hospitality design guides.

Axel Venn曾在埃森福克旺设计学院学习设计和自由作曲，他现在在柏林生活并开展研究工作，并担任一些国际公司的创意和科学顾问。Axel Venn教授已出版了25本关于色彩学、市场营销和潮流风格，以及实验学科的书。迄今为止，他的书已被翻译成12种语言。

Janina Venn-Rosky毕业于汉堡美术大学（UFBK），为设计专业。作为共同作者，她至今已出版了7本书和15本顶级配色教程。Janina Venn-Rosky和丈夫社会经济学硕士Alexander Venn与Axel Venn多年来有着密切的合作，尤其是在实证研究和实践相关的应用领域。

Corinna Kretschmar-Joehnk，JOI-Design GmbH的联席董事总经理，是酒店室内建筑和设计业内顶尖的专家之一。她曾在维尔茨堡学习艺术史，并于德特莫尔德学习室内设计。20多年来，在两大酒店品牌和独立精品酒店的建设上，Corinna富有远见的理念深得顾客喜爱，并引领了全球酒店潮流。自2003年以来，她和丈夫Peter Joehnk，带领JOI-Design成为欧洲领先的室内建筑工作室。夫妇两人曾在之前出版的酒店指南设计上分享了他们的创作激情。

INDEX OF PRODUCTS   产品索引

Company	Term	Chap.	Product	Copyright photo	Page	Position
JAB Anstoetz	Introduction	0	Charme Collection	JAB Anstoetz	9	right
JAB Anstoetz	luxurious	I	CHIVASSO, Exposure Collection	JAB Anstoetz	24	centre
JAB Anstoetz	luxurious	I	CHIVASSO wallpapers, Wonderwalls Collection	JAB Anstoetz	24	bottom
Kvadrat A/S	luxurious	I	Seating furniture fabric: Sudden 003, colour: silver	JOI-Design	24	centre left
DELIUS GmbH	luxurious	I	Fabric Modesta, Scala Collection	DELIUS	26	top right
DELIUS GmbH	luxurious	I	Fabric Adara, Scala Collection	DELIUS	26	centre left
ege® GmbH	luxurious	I	Carpet: ege® Highline 1400 80/20	JOI-Design	28	top left
ege® GmbH	luxurious	I	Carpet: ege® Highline 1400 80/20	JOI-Design	29	
Ehrlich Leder GmbH	elegant	I	Longtime Nappa sierra and Longtime Nappa hasel	Ehrlich Leder GmbH	40	bottom left
Voglauer	elegant	I	Interior works Hilton Frankfurt Airport Hotel	JOI-Design	40+41	
ABB Asea Brown Boveri Ltd. Busch-Jaeger Elektro GmbH	elegant	I	Decento® range, colour: studio white	ABB Asea Brown Boveri Ltd. Busch-Jaeger Elektro GmbH	42	top left
Panaz	sophisticated	I	Bedhead fabric: Sheema	JOI-Design	54	top left
ege® GmbH	sophisticated	I	Carpet: ege® Highline 1400 80/20	JOI-Design	54	centre left
ege® GmbH	sophisticated	I	Carpet: ege® Highline 1400 80/20	JOI-Design	55	
DELIUS GmbH	sophisticated	I	Curtain fabric, Edward DIMOUT	JOI-Design	55	
ege® GmbH	sophisticated	I	Carpet: ege® Highline 1400 80/20	JOI-Design	56	bottom left
Panaz	sophisticated	I	Bedhead fabric: Sheema	JOI-Design	56	bottom centre
Ehrlich Leder GmbH	sophisticated	I	Imperial Emotions	Ehrlich Leder GmbH	56	top right
ege® GmbH	sophisticated	I	Carpet: ege® Highline 1400 80/20	Christoph Kraneburg, Köln	57	top left
Voglauer	stylish	I	Interior works Hilton Frankfurt Airport Hotel	JOI-Design	68–71	
JAB Anstoetz	stylish	I	Marzano	JAB Anstoetz	71	bottom centre
FSB Franz Schneider Brakel GmbH + Co KG	stylish	I	FSB 72 1163, bronze bright patinated, door handle design: Hans Kollhoff, product range: Bronze fitting collection	FSB	71	top left
JAB Anstoetz	world-class	I	Soleil Bleu, Salon Collection	JAB Anstoetz	82	top left
JAB Anstoetz	world-class	I	CHIVASSO Wallpapers, Whispers Collection	JAB Anstoetz	82	centre
Panaz	world-class	I	Decorative cushion covers: Imperial Velvet, colour: 129 ocean azure	JOI-Design	82	bottom left
Panaz	world-class	I	Decorative cushion covers: Imperial Velvet, colour: 129 ocean azure	Dom Hotel Köln	83	
Panaz	world-class	I	Curtain: Taffeta Blackout Collection, colour: 814 Espresso	Dom Hotel Köln	83	
Kvadrat A/S	world-class	I	Upholstery: Remix 2 by Giulio Ridolfo	Kvadrat	84	bottom left
ROLF BENZ	world-class	I	Longchair, model 280, leather cover, polished chrome frame, design: Böhme/Reiter	ROLF BENZ	84+85	top centre
JAB Anstoetz	world-class	I	Kavallerietuch Collection	JAB Anstoetz	86	top left
Panaz	world-class	I	Opulence Collection with Enchantment, Adore, Glory, and Imperial Velvet	Panaz	86	bottom left
Panaz	world-class	I	Opulence Collection with Glory, Cherish, Enchantment, Splendour, and Imperial Velvet	Panaz	86+87	
DELIUS GmbH	top-notch	I	Fabric: Rondo, Scala Collection	DELIUS	98	top left
ege® GmbH	top-notch	I	Carpet: ege® Highline 1400 80/20	JOI-Design	99	
Voglauer	top-notch	I	Interior works Hilton Frankfurt Airport Hotel	JOI-Design	99	
Kvadrat A/S	top-notch	I	Seating furniture fabric: Sudden 003, colour: silver	JOI-Design	101	right
ABB Asea Brown Boveri Ltd. Busch-Jaeger Elektro GmbH	top-notch	I	Product range: alpha exclusive®, colour: platin	ABB Asea Brown Boveri Ltd. Busch-Jaeger Elektro GmbH	101	bottom left
JAB Anstoetz	neat	II	UV-Pro-Outdoor Collection	JAB Anstoetz	114	top left
JAB Anstoetz	neat	II	CHIVASSO, Whispers Collection	JAB Anstoetz	114	bottom right
Kvadrat A/S	neat	II	Upholstery: Canvas by Giulio Ridolfo	Kvadrat	114	centre

Company	Term	Chap.	Product	Copyright photo	Page	Position
DELIUS GmbH	neat	II	Fabrics: Maxim and Dimout	InterContinental Resort Berchtesgarden	115	
JAB Anstoetz carpets	neat	II	Noblesse Collection	JAB Anstoetz	116	right, centre
DELIUS GmbH	neat	II	Fabric: Juno, Prisma Collection	DELIUS	116	centre left
Voglauer	neat	II	Interior finish Thermenhotel Stoiser	Leuchtende Hotel Fotografie	116	top + bottom left; bottom centre
ROLF BENZ	neat	II	Chair 275, leather cover, walnut frame, design: Zeichen und Wunder	ROLF BENZ	117	
Kvadrat A/S	rejuvenating	II	Curtain fabric: Etzel (by Fanny Aronsen) with STOPLIGHT coating, colour: green-beige	JOI-Design	128	bottom left
ege® GmbH	rejuvenating	II	Carpet: ege® Highline 1400 80/20	JOI-Design	129	
Kvadrat A/S	rejuvenating	II	Curtain fabric: Etzel (by Fanny Aronsen) with STOPLIGHT coating, colour: green-beige	JOI-Design	129	
DELIUS GmbH	rejuvenating	II	Screen curtain fabric: Piera DELILIGHT	JOI-Design	130	bottom centre
Panaz	rejuvenating	II	Bedhead fabric: Sheema	JOI-Design	131	right
JAB Anstoetz	relaxing	II	Soleil Bleu, Salon Collection	JAB Anstoetz	142	centre
JAB Anstoetz	relaxing	II	Atmospheric picture	JAB Anstoetz	142	bottom left
JAB Anstoetz	relaxing	II	KeyWest Collection	JAB Anstoetz	142	bottom centre
ege® GmbH	relaxing	II	Carpet: ege® Highline 910	Melanie Schmidt	142	top left
JAB Anstoetz carpets	relaxing	II	Spotline Collection	JAB Anstoetz	143	
JAB Anstoetz	relaxing	II	UV-Pro-Outdoor Collection	JAB Anstoetz	144	bottom left; top centre
JAB Anstoetz	relaxing	II	UV-Pro-Outdoor Collection	JAB Anstoetz	145	top left
Panaz	relaxing	II	Lounge chairs fabric: Linear, colour: 226 lime	JOI-Design	146	top left
ege® GmbH	relaxing	II	Carpet: ege® soft dreams	JOI-Design	146+147	
Panaz	relaxing	II	Sofa fabric: Linear, colour: 213 aubergine	JOI-Design	146+147	
JAB Anstoetz	vitalising	II	Soleil Bleu, Tropicana Collection	JAB Anstoetz	158	bottom left
JAB Anstoetz	vitalising	II	Lounge Collection	JAB Anstoetz	158	centre
JAB Anstoetz	vitalising	II	CARLUCCI, Exposure Collection	JAB Anstoetz	160	top left
JAB Anstoetz	vitalising	II	Colour paint Collection	JAB Anstoetz	160	bottom left
JAB Anstoetz	vitalising	II	CARLUCCI, Exposure Collection	JAB Anstoetz	160	top right
JAB Anstoetz	vitalising	II	Soleil Bleu, Tropicana Collection	JAB Anstoetz	160	bottom centre
JAB Anstoetz	vitalising	II	Soleil Bleu, Tropicana Collection	JAB Anstoetz	161	right
Kvadrat A/S	authentic	III	Fabric: Molly 143	JOI-Design	174	top left
ROLF BENZ	authentic	III	Plaid with cable stitch, 100 % superfine lambswool	ROLF BENZ	174	bottom right
JAB Anstoetz	authentic	III	CARLUCCI, Exposure Collection	JAB Anstoetz	176	top centre
Ehrlich Leder GmbH	authentic	III	Sleek leather	Ehrlich Leder GmbH	177	bottom left
Kvadrat A/S	authentic	III	Upholstery Collection by Alfredo Häberli: Plot:	Kvadrat	178	centre
Kvadrat A/S	authentic	III	Seating furniture fabric: Fanny Aronson Sunniva 717	Kvadrat	178+179	
Kvadrat A/S	restrained	III	Seating furniture fabric: Happy 106	JOI-Design	190	top left
JAB Anstoetz	restrained	III	CHIVASSO, Whispers Collection	JAB Anstoetz	191	bottom right
JAB Anstoetz	restrained	III	CHIVASSO, Whispers Collection	JAB Anstoetz	192	centre
ege® GmbH	restrained	III	Carpet: ege® Highline 1100	Christoph Kraneburg, Köln	192	top left
ege® GmbH	restrained	III	Carpet: ege® Highline 910	Christoph Kraneburg, Köln	192	bottom left
ABB Asea Brown Boveri Ltd. Busch-Jaeger Elektro GmbH	restrained	III	Product range: Busch-axcent®, colour: white glass	ABB Asea Brown Boveri Ltd. Busch-Jaeger Elektro GmbH	192+193	bottom centre
JAB Anstoetz	restrained	III	CHIVASSO, Whispers Collection	JAB Anstoetz	192+193	top centre

Company	Term	Chap.	Product	Copyright photo	Page	Position
ege® GmbH	restrained	III	Carpet: ege® Highline 1100	Christoph Kraneburg, Köln	193	right
ege® GmbH	family-friendly	III	Carpet: ege® Highline 1100	ege®	204	bottom
Kvadrat A/S	family-friendly	III	Upholstery: Remix by Giulio Ridolfo	Kvadrat	204	centre right
Ehrlich Leder GmbH	family-friendly	III	Royal Nappa colour block	Ehrlich Leder GmbH	204	top left
ege® GmbH	family-friendly	III	Carpet: ege® Highline 1100	ege®	205	right
FSB Franz Schneider Brakel GmbH + Co KG	puristic	III	FSB 25 1035, isis T300 electronic access management, door handle design: Heike Falkenberg, product range: isis	FSB	216	top left
FSB Franz Schneider Brakel GmbH + Co KG	puristic	III	FSB 82 8224, drop-down support rails, design: Hartmut Weise, product range: Barrier-free ErgoSystem®	FSB	216	bottom left
ROLF BENZ	puristic	III	Armchair model 266, fabric cover, with star-shaped steel base, matt chrome surface, design: Cuno Frommherz	ROLF BENZ	216	bottom right
Ehrlich Leder GmbH	puristic	III	Longtime Nappa papyrus	Ehrlich Leder GmbH	216	centre
Voglauer	puristic	III	Interior works Radisson BLU Hotel Amsterdam Airport	Arne Jennard	217	
Voglauer	puristic	III	Interior works Radisson BLU Hotel Amsterdam Airport	Arne Jennard	218	top left
ROLF BENZ	puristic	III	Armchair model 266, leather cover, base plate with 360 degree swivel mechanism, design: Cuno Frommherz	ROLF BENZ	218	top centre
JAB Anstoetz	puristic	III	Black & White Collection	JAB Anstoetz	219	bottom left
ABB Asea Brown Boveri Ltd. Busch-Jaeger Elektro GmbH	puristic	III	Product range: Busch-axcent®, colour: maison-beige	ABB Asea Brown Boveri Ltd. Busch-Jaeger Elektro GmbH	219	bootm right
DELIUS GmbH	historic	III	New Velvets Collection	DELIUS	230	top left
DELIUS GmbH	historic	III	Fabric: Rondo, Scala Collection	DELIUS	230	bottom left
DELIUS GmbH	historic	III	Storia Collection	DELIUS	230	centre
DELIUS GmbH	historic	III	New Velvets Collection	DELIUS	230	bottom right
DELIUS GmbH	historic	III	Storia Collection	DELIUS	231	
DELIUS GmbH	historic	III	Picada DELIGARD	DELIUS	232	top left
Kvadrat A/S	historic	III	Curtain Collection by Cristian Zuzunaga: Castillo	Kvadrat	233	bottom left
Panaz	historic	III	Article: Imperial Velvet, colour: 400 red; cushions in colour 702 chocolate and 807 mushroom	Panaz	233	right
ABB Asea Brown Boveri Ltd. Busch-Jaeger Elektro GmbH	historic	III	Product range: carat®, colour bronze	ABB Asea Brown Boveri Ltd. Busch-Jaeger Elektro GmbH	233	top left
JAB Anstoetz	romantic	IV	Best of Structures Collection	JAB Anstoetz	248	top left
Panaz	romantic	IV	Seating furniture fabric: Mohair	Panaz	248	bottom left
DELIUS GmbH	romantic	IV	Prisma Collection	DELIUS	248	centre
Panaz	romantic	IV	Fabric: Indulgence Velvet, colour: 624	Panaz	249	
JAB Anstoetz	poetic	IV	Outdoor Collection	JAB Anstoetz	260	top left
JAB Anstoetz	poetic	IV	Four Seasons Collection	JAB Anstoetz	260	bottom left
JAB Anstoetz	poetic	IV	CHIVASSO, Whispers Collection	JAB Anstoetz	260	centre
JAB Anstoetz	poetic	IV	Soleil Bleu, Collections: Salon & Provence	JAB Anstoetz	261	
JAB Anstoetz	poetic	IV	Soleil Bleu, Salon Collection	JAB Anstoetz	261	top centre
JAB Anstoetz	imaginative	IV	Soleil Bleu, Stella Celerina Collection	JAB Anstoetz	274	bottom right
Panaz	imaginative	IV	Bedhead fabric: Sheema	JOI-Design	275	
ege® GmbH	inspiring	IV	Carpet: ege® texture	JOI-Design	287	
JAB Anstoetz	classic	IV	Red Desire Collection	JAB Anstoetz	298	bottom right
JAB Anstoetz	classic	IV	Black & White Collection	JAB Anstoetz	298	centre
JAB Anstoetz	classic	IV	Atmospheric picture	JAB Anstoetz	300	bottom left
DELIUS GmbH	snug	IV	New Velvets Collection	DELIUS	314	bottom centre
Voglauer	snug	IV	Interior works Fährhaus Munkmarsch	Ydo Sol	314	centre left

Company	Term	Chap.	Product	Copyright photo	Page	Position
Voglauer	snug	IV	Interior works Fährhaus Munkmarsch	JOI-Design	314	top + centre left
Voglauer	snug	IV	Interior works Fährhaus Munkmarsch	Ydo Sol	315	
Ehrlich Leder GmbH	snug	IV	Jumbo Nubuk nature	Ehrlich Leder GmbH	316	top right
DELIUS GmbH	snug	IV	Fabrics: Mirano and Juno, Prisma Collection	DELIUS	317	top left
JAB Anstoetz	nonconformist	V	CARLUCCI, Exposure Collection	JAB Anstoetz	330	centre
Panaz	nonconformist	V	Aston with additional acrylic shade surfaces Protection	Panaz	330	bottom left
Voglauer hotel concept	nonconformist	V	Design & concept	Voglauer hotel concept	330	top left
Voglauer hotel concept	nonconformist	V	Design & concept	Voglauer hotel concept	331	
ege® GmbH	bohemian	V	Carpet: ege® Highline 1400 80/20	JOI-Design	343	
Kvadrat A/S	bohemian	V	Curtain fabric: Checkpoint 193	JOI-Design	343	
JAB Anstoetz	chic	V	Charme Collection	JAB Anstoetz	354	bottom right
Kvadrat A/S	chic	V	Upholstery fabric	Kvadrat	356	
Kvadrat A/S	chic	V	Upholstery by Giulio Ridolfo: Plot	Kvadrat	356+357	top centre
Kvadrat A/S	chic	V	Curtain fabric: Lux	Kvadrat	357	right
Ehrlich Leder GmbH	chic	V	Bristol Nappa kardinal	Ehrlich Leder GmbH	357	bottom left
Voglauer	chic	V	Interior works Le Méridien Brüssel	Lutz Voderwühlbecke	358+359	
ABB Asea Brown Boveri Ltd. Busch-Jaeger Elektro GmbH	futuristic	V	Control element Busch-priOn®, colour: corian studio white	ABB Asea Brown Boveri Ltd. Busch-Jaeger Elektro GmbH	382	top right
ABB Asea Brown Boveri Ltd. Busch-Jaeger Elektro GmbH	futuristic	V	Control element Busch-triton®, colour: studio white high glossy	ABB Asea Brown Boveri Ltd. Busch-Jaeger Elektro GmbH	382	bottom right
Kvadrat A/S	futuristic	V	Fabric chair cushion: Divina	JOI-Design	383	
JAB Anstoetz	charming	V	CHIVASSO, Whispers Collection	JAB Anstoetz	396	top; bottom left
JAB Anstoetz	charming	V	Four Seasons Collection	JAB Anstoetz	396	centre; bottom right
JAB Anstoetz	charming	V	Four Seasons Collection	JAB Anstoetz	397	
JAB Anstoetz	Oriental	VI	Marrakesch Collections	JAB Anstoetz	410	top + bottom left; bottom right
DELIUS GmbH	Oriental	VI	Upholstery fabric: Ruben	JOI-Design	413	right
JAB Anstoetz	African	VI	Soleil Bleu, Lodge Collection	JAB Anstoetz	414	bottom right

INDEX OF HOTELS AND DESIGNS
酒店设计检索

Hotel / 酒店	Term / 检索词	Chapter / 章节	Page / 页码	Position / 位置	Copyright / 版权
Hilton Frankfurt Airport Hotel	Introduction / 引言		8	right / 右	JOI-Design
Hilton Munich Park Hotel	luxurious / 豪情奢华	I	24–26		JOI-Design
Le Clervaux Boutique & Design Hotel	luxurious / 豪情奢华	I	27, 28 + 29		JOI-Design
Hilton Frankfurt Airport Hotel	elegant / 优雅内敛	I	40 + 41		JOI-Design
Steigenberger Grandhotel & Spa Heringsdorf	elegant / 优雅内敛	I	42 + 43		JOI-Design

Each hotel is different and unique in its own way, due to its classification, its location, its orientation and – as a result of the aforementioned criteria – its design.
Here, the design of the **Hilton Munich Park Hotel** stands for the term "luxurious", because, from the lobby to the bar and the rooms and suites, the colour theme is in perfect harmony with the entire interior, and a particularly warm atmosphere prevails despite the exclusiveness. Formally, the adjoining English Garden – with its blossom and leaf motifs – was a source of inspiration for the design concept (interior design: JOI-Design).

Finalist: European Hospitality Award 2012

酒店因不同的类别、地理位置、朝向而拥有不同的设计，每个都独一无二。其中，慕尼黑公园希尔顿酒店（Hilton Munich Park Hotel）的设计代表了所谓的"奢华"，从大堂到酒吧、客房和套房，它的色彩主题和整个室内完美契合。作为高档酒店，它却拥有十分温馨的氛围。从形式上看，附近英式花园中花花叶叶的形状成为设计理念的灵感源泉（室内设计：JOI-Design）。
入围：2012年欧洲酒店大奖。

Despite an apparently far more exciting design – a modern interpretation of opulent baroque motifs and shapes in the carpet design, wallpaper and furniture – the Le Clervaux Boutique & Design Hotel is perfectly integrated into the luxurious colour world, too. This clarifies the diversity of possible interpretations. The view of the neighbouring castle became the guiding theme in the design for a sophisticated mixture of old and new (interior design: JOI-Design).

Boutique Media: Boutique Design Award / Most surprising visual element 2012. Heinze Verlag: Heinze Publikums Preis 2013

勒克莱沃精品设计酒店（Le Clervaux Boutique & Design Hotel）的设计显然更令人振奋，它的地毯设计、墙纸和家具对巴洛克风格华丽的图案和形状进行了现代诠释。酒店同时完美地融入了奢华的色彩世界。这表明，诠释方式有多种可能。作为新旧精心融合的设计，指导原则是可将邻近城堡的风景一览无遗（室内设计：JOI-Design）。
曾获2012年精品酒店设计大奖 / 最令人惊艳的视觉元素奖；并刊载于海因策出版社（Heinze Verlag）的《Heinze Publikums Preis 2013》。

Polite restraint, paired with cool elegance and a galvanising haptic excitement generated by the use of contrasting materials like glass and steel combined with exquisite leather and delicate fabrics. This is what guests find in the **Hilton Frankfurt Airport Hotel**. The design concept accepts the balancing act and conceives interiors that reflect the dynamics of the building at the airport and convey warmth nonetheless (interior design: JOI-Design).

Top hotel: Top hotel opening award 2011 – Kategorie ‚Luxus'. International Hotel Awards: Best New Hotel Construction & Design Germany 2012

法兰克福机场希尔顿酒店（Hilton Frankfurt Airport Hotel）有着文雅而低调的特质，配以冷色调营造的优雅，通过对比的玻璃和钢材质，结合精美的皮革和精致的面料，为客人带来兴奋感。其设计理念为室内设计应于人们行为和感知中寻求平衡，它既能反映机场建筑的动感，又能传递温暖（室内设计：JOI-Design）。
2011年顶级酒店大奖：奢华类高级酒店开业奖；2012年国际酒店大奖：德国最佳新酒店建筑和设计奖。

This contrasts with the design of the **Steigenberger Grandhotel & Spa Heringsdorf** on the island of Usedom in the Baltic Sea: here, an elegant colour range meets rooms that radiate elegance and warmth generated by soft shapes, colours and materials. The ambience radiates "casual maritime chic"; its fine details are reminiscent of the days of the former imperial bath (interior design: JOI-Design).

与此相反，位于波罗的海乌泽多姆岛屿上的黑林格斯多尔夫斯泰根博格温泉大酒店（Steigenberger Grandhotel & Spa Heringsdorf），在设计上通过柔和的外观、色彩和材料打造了优雅、温馨的客房，而"优雅内敛"风格的色彩与客房的优雅、温馨相协调。它营造出"于海上休闲、潇洒生活"的氛围；精致的细节让人想起了帝国时代沐浴的时光（室内设计：JOI设计）。

Le Clervaux Boutique & Design Hotel	sophisticated / 精致得体	I	54	centre left; centre / 左中；中	JOI-Design
Hotel Gendarm nouveau	sophisticated / 精致得体	I	54	top left; bottom centre / 左上；中下	JOI-Design
Le Clervaux Boutique & Design Hotel	sophisticated / 精致得体	I	55		JOI-Design
Le Clervaux Boutique & Design Hotel	sophisticated / 精致得体	I	56	top and bottom left / 左上；左下	JOI-Design
Le Clervaux Boutique & Design Hotel	sophisticated / 精致得体	I	57	top left / 左上	Christoph Kraneburg, Köln
Le Clervaux Boutique & Design Hotel	sophisticated / 精致得体	I	57	right / 右	JOI-Design
Hotel Gendarm nouveau	sophisticated / 精致得体	I	56 + 57	bottom centre / 中下	JOI-Design
Hilton Frankfurt Airport Hotel	stylish / 时尚现代	I	68 + 69, 70 + 71		JOI-Design

Arrive and feel at ease in a young, sophisticated ambience: rich colours combined with elegant black and grey, the decor is striking and simultaneously gorgeously homely and very modern. These are the characteristics of the designs with corresponding colour concepts of the **Le Clervaux Boutique & Design Hotel** as well as the **Hotel Gendarm nouveau** in Berlin. Situated directly at the Gendarmenmarkt, the interior decoration takes up historical quotations and transports them to the modern era (interior design: JOI-Design).

勒克莱沃精品设计酒店（Le Clervaux Boutique & Design Hotel）和新皇家卫士酒店（Hotel Gendarm nouveau）坐落于柏林的御林广场，丰富的色彩与典雅的黑色和灰色相结合，令人惊叹不已的华丽内饰营造出亲切而又现代的氛围。它们年轻而精致，客人很快感受到这一点。室内也采用了相应的色彩理念，装修则呼应了御林广场的历史渊源，但不失现代（室内设计：JOI-Design）。

The Presidential Suite in the **Hilton Frankfurt Airport Hotel**: the luxurious sleeping area is designed in noble silver-grey shades, and small chandeliers above the bedside tables immerse the room in sublime light. The freestanding black jacuzzi bathtub was purpose-built by an artist with self-levelling coating and framed metal strips. The mirror with a mother-of-pearl frame as well as the gold-black granite also emphasise the consummate style of the suite. The interior design radiates "sophisticated internationality", which is ideal for the airport location and leaves nothing to be desired (interior design: JOI-Design).

Top hotel: Top hotel opening award 2011 – Kategorie ‚Luxus'. International Hotel Awards: Best New Hotel Construction & Design Germany 2012

法兰克福机场希尔顿酒店（Hilton Frankfurt Airport Hotel）总统套房：奢华的睡眠区采用高贵的银灰色调设计，床头柜上的小吊灯使得房间笼罩在庄严的灯光下。黑色的独立式按摩浴缸是艺术家用自流平涂料和金属框架条特制的，而以珍珠母贝镶框的镜子和金黑色花岗岩将套房的美发挥到极致。室内设计"既有国际范儿又精致"，对于机场的位置来说非常理想，不需再补充什么（室内设计：JOI-Design）。
2011年顶级酒店大奖：奢华类高级酒店开业奖；2012年国际酒店大奖：德国最佳新酒店建筑和设计奖。

Dom Hotel Köln	world-class / 世界一流	I	82 + 83		Dom Hotel Köln
Le Méridien Grand Hotel Nürnberg	world-class / 世界一流	I	84		Le Méridien Grand Hotel Nürnberg
Le Méridien Grand Hotel Nürnberg	world-class / 世界一流	I	85		JOI-Design
Showroom	world-class / 世界一流	I	86 + 87		Panaz
Hilton Vienna Danube	top-notch / 首屈一指	I	98	bottom left; centre/左下；中	JOI-Design
Hilton Frankfurt Airport Hotel	top-notch / 首屈一指	I	98 + 99		JOI-Design
Hilton Munich Park	top-notch / 首屈一指	I	100 + 101		JOI-Design

The room design in the **Dom Hotel Köln** is as world-class as the view to the cathedral, which guests can enjoy from many of the rooms. Gentle shades of beige and brown, accentuated by decorative cushions on the blue-turquoise spectrum, generate a harmonious interior impression, which does without frills and instead impresses with mostly straight or slightly curved shapes. Seating furniture with high-quality leather upholstery and a similarly covered headboard on the bed, which reaches up to the ceiling, gives the ambience the final touch. The design for the rooms in the former "Blau-Gold-Haus" (Eau de Cologne) is in terms of colour and form reminiscent of the façade and its construction in the 1950s (interior design: JOI-Design).

科隆多姆酒店（Dom Hotel Köln）的房间设计和科隆大教堂一样，均属世界一流，多间房内都可以看到科隆大教堂的美景。由蓝绿松石色的装饰性靠垫凸显温和的米色和棕色色调，给人以和谐的室内印象；并没有多余的装饰，而室内大多直或稍弯曲的形状给人留下深刻的印象。座椅采用优质真皮座套，床上则采用了同样用优质真皮包裹的床头板，它延伸至天花板处，成为点睛之笔。"科隆香水客房"因其色彩和造型使人联想到20世纪50年代的科隆建筑和外观而得名（室内设计：JOI-Design）。

The rooms and suites in the **Le Méridien Grand Hotel Nürnberg** are equally "world-class", but overall they feature a younger design with delicate accentuations in red and cream, in combination with classic black-and-white, while being similarly unadorned. Large-format carpet and wall patterns further animate the interiors. The design concept is reminiscent of the brand's French roots – the hound's-tooth check reminds one of Coco Chanel or the suitcase details of Louis Vuitton (interior design: JOI-Design).

纽伦堡艾芙大酒店（Le Méridien Grand Hotel Nürnberg）的客房和套房同样"世界一流"，总体来说，它们的设计朴实但较年轻化，结合经典的黑色和白色，精致的红色色调和奶油色调突出。通过大面积的地毯和墙上的图案活跃了室内的氛围。千鸟格图案易使人联想起香奈儿或LV的旅行箱，这种设计使人联想到该品牌的法国根源（室内设计：JOI设计）。

A third facet of the term "world-class" is illustrated by the **showroom** on the next double page – there is no sign of restraint anymore: first-rate fabrics in vibrant shades of red with various nuances and small as well as large-format patterns pervade the scenery (Panaz).

下个跨页的陈列室（Showroom）充分阐明了"世界一流"一词的又一方面，在这里，设计不再节制，一流的面料、充满活力的各种红色色调和大面积的图案随处可见（Panaz）。

The **Hilton Vienna Danube** welcomes its guests in a two-storey lobby, where a collection of several spherical glass hanging lamps that are suspended from the ceiling complement the height of the room. Wide, ceiling-high columns flank the aisle and direct the gaze to a staircase leading to the next level. A carpet in subtle beige-red with stripes interrupted by an abstract floral pattern encapsulates with the colours and shapes of the "top-notch" in the same way as the reception area. The renovations focused on an unrestricted view of the Danube flowing past, and the theme of moving waves runs through the entire design concept (interior design: JOI-Design).

维也纳希尔顿多瑙河酒店（Hilton Vienna Danube）拥有一个两层高的大堂，天花板上悬挂着许多球状玻璃吊灯，可从视觉上调和高度。高至天花的宽柱子列至旁边的过道，引导视线至通往另一层的楼梯。淡淡的米色红条纹地毯视觉上则以与接待区同系列、色彩和形状都堪称世界一流的花卉抽象画为点缀。翻修焦点在于确保多瑙河风景的一览无遗，涟漪、轻波的主题也贯穿于整个设计概念（室内设计：JOI-Design）。

The ballroom "Globe" in the **Hilton Frankfurt Airport Hotel** represents "top-notch" colour design in public hotel areas: walls inlaid with Swarovski crystals, which seem to "flash" through the room like luminous strips, together with wave-shaped cove lighting, provides for more than impressive lighting effects. Warm shades of red and gentle beige-brown hues as well as flowing shapes in the carpet design emphasise the entire ambience. The ballroom as a "building inside the building" repeats the external cubature of the building in the interior and uses futuristic elements to build a bridge to the airport location (interior design: JOI-Design).

Top hotel: Top hotel opening award 2011 – Kategorie ‚Luxus'. International Hotel Awards: Best New Hotel Construction & Design Germany 2012

法兰克福机场希尔顿酒店（Hilton Frankfurt Airport Hotel）的"环球"舞厅是酒店公共区域色彩设计的"一流"典范：墙壁上镶嵌着施华洛世奇水晶，在房间中熠熠生辉，它们与波浪形的凹圆形天棚照明一起，产生的照明效果令人印象更深刻。温暖的红色色调和柔和的米黄褐色调，以及地毯上的流动形状突出了整个氛围。舞厅作为"建筑内部的建筑"，在室内重复了建筑外部的求和容积法，通过未来主义的元素，呼应了机场的地理位置（室内设计：JOI-Design）。
2011年顶级酒店大奖：奢华类高级酒店开业奖；2012年国际酒店大奖：德国最佳新酒店建筑和设计奖。

Delicate patterns, exquisite materials, warm colours, and mainly soft shapes, highlighted by a combination of smooth, fluffy and shiny fabrics, can also be found in the design of the suites and bars of the **Hilton Munich Park Hotel**, thus constituting another excellent example of a hotel in the world of "top-notch" colour. The adjoining park remains visible as a design principle in all areas and adds accentuations with delicate leaf and blossom motifs (interior design: JOI-Design).

Finalist: European Hospitality Award 2012

慕尼黑公园希尔顿酒店（Hilton Munich Park Hotel）套房和酒吧通过或光滑或蓬松而有光泽的面料组合，凸显了精美的花纹、考究的用料、温暖的色调和以柔和为主的形状，成为了一流酒店色彩设计的极好例子。各个区域将毗邻的公园仍然可见作为设计原则，并辅以花花叶叶的精美图案（室内设计：JOI-Design）。
入围：2012年欧洲酒店大奖。

InterContinental Berchtesgarden Resort	neat / 简洁淡雅	II	114 + 115	centre left / 左中	InterContinental Berchtesgarden Resort
Thermenhotel Stoiser	neat / 简洁淡雅	II	116	top and bottom left; bottom centre / 左上，左下；左中	Leuchtende Hotel Fotografie
Hilton Frankfurt Airport Hotel	rejuvenating / 焕发新生	II	128 + 129		JOI-Design
Steigenberger Grandhotel & Spa Heringsdorf	rejuvenating / 焕发新生	II	130	top, centre and bottom left / 上，中，左下	JOI-Design
Mövenpick Hotel Munich Airport	rejuvenating / 焕发新生	II	130 + 131	centre bottom, right / 中下，右	JOI-Design

Clear lines, reduced patterns, natural colours like shades of brown, beige and soft yellow characterise the design of the rooms in the **InterContinental Berchtesgarden Resort** and generate an extremely neat interior atmosphere, which invites guests to relax. This language of form and colour, accentuated with subtle blue curtains, decorative cushions and upholstery fabrics, also characterises the appearance of rooms in the **Thermenhotel Stoiser,** which is slightly straighter in its design vocabulary (interior design: Dipl. Ing. Strohecker ZT GmbH).

贝希特斯加登洲际度假酒店（InterContinental Berchtesgarden Resort）的客房设计使用了干净的线条、简化的图案，以及棕色、米色和柔和的黄色色调等自然的色彩，营造出一个非常整洁的室内氛围，欢迎客人来此放松。淡蓝色的窗帘、装饰性靠垫和装饰面料更凸显了这种形态和色彩风格。**特蒙斯特宜赛尔酒店**（Thermenhotel Stoiser）的外观也属于这种风格，但宜赛尔酒店风格更明显（室内设计：Dipl. Ing. Strohecker ZT GmbH）。

The **Hilton Frankfurt Airport Hotel** is situated at the infrastructural hub of the Main metropolis, which is why pure regeneration and deceleration are given priority. Surrounded by warm beige hues and fresh green, the guest can leave behind the noisy traffic environment and find peace and quiet. In a very subtle way, the wavy lines of the carpet and the abstracted motifs of the pictures take up symbols of mobility, which allude to the location. The bright bathrooms with their fresh and warm colour emphases are also conducive to relaxation (interior design: JOI-Design).

Top hotel: Top hotel opening award 2011 – Kategorie „Luxus". International Hotel Awards: Best New Hotel Construction & Design Germany 2012

法兰克福机场希尔顿酒店（Hilton Frankfurt Airport Hotel）坐落于机场基础设施的中心，酒店的再生能力和帮助客人放慢步伐的能力成为建筑和设计的重中之重。为温暖的米色色调以及新鲜的绿色色调所环绕，客人可以在离开嘈杂的交通环境后寻求和平与宁静。地毯上的波浪线和壁画上抽象的图案微妙地象征了流动性，暗示了酒店所处的位置。浴室非常明亮，且主要使用新鲜和温暖的色彩，有利于客人放松（室内设计：JOI-Design）。
2011年顶级酒店大奖：奢华类高级酒店开业奖；2012年国际酒店大奖：德国最佳新酒店建筑和设计奖。

A similar concept stands behind the **Mövenpick Hotel Munich Airport**: as an antipole to the technology of the nearby airport, nature served as a decisive source of inspiration in order to create a soothing contrast. Pine-green walls are next to sand-coloured tiles and warm walnut wood shades. Here, the green curtains and decorative cushions in the rooms are even a little fresher than in the Hilton Frankfurt Airport Hotel, whereby the designs of the carpets and curtains generally feature more reduced patterns (interior design: JOI-Design).

Adex Silver Award 2009

慕尼黑机场莫凡彼酒店（Mövenpick Hotel Munich Airport）也采用了类似的理念，与附近机场给人的技术感恰恰相反，自然，成为了酒店设计的灵感源泉，以营造能抚慰人的反差环境。酒店采用了松绿色的墙壁、沙色的瓷砖和温暖的胡桃木色调。与法兰克福机场希尔顿酒店相比，莫凡彼酒店房间内的绿色窗帘和靠垫装饰更清新，设计地毯和窗帘时也采用了更简化的图案（室内设计：JOI-Design）。
曾获2009年Adex大奖银奖。

A rejuvenating approach in the design vocabulary is pursued not only in many business hotels but also in many hotels with a focus on recreational holidays and wellness. These include the **Steigenberger Grandhotel & Spa Heringsdorf.** In the pool and sauna area as well as in the relaxation rooms, the design plays with indirect lighting; warm beige and orange hues in combination with natural shapes and materials dominate the interiors (interior design: JOI-Design).

许多商务酒店和比较注重休闲度假和健康的酒店都追求可恢复活力的复原型设计手法，如黑林格斯多尔夫斯泰根博格温泉大酒店（Steigenberger Grandhotel & Spa Heringsdorf）。它的游泳池和桑拿区，以及休息室，设计了许多活泼的间接照明；室内则以与天然形状和材料结合的米色、橙色等温暖色调为主宰（室内设计：JOI-Design）。

Landgasthof Bären	relaxing / 轻松自在	II	142	top left / 左上	Melanie Schmidt
Steigenberger Grandhotel & Spa Heringsdorf	relaxing / 轻松自在	II	144	centre left; bottom centre / 左中；中下	JOI-Design
Hotel Ritter Durbach	relaxing / 轻松自在	II	144	top left / 左上	JOI-Design
Mövenpick Hotel Munich Airport	relaxing / 轻松自在	II	145	right / 右	JOI-Design
Hotel Ritter Durbach	relaxing / 轻松自在	II	146 + 147		JOI-Design
Park Inn Nürnberg	vitalising / 无限活力	II	158 + 159		JOI-Design

Relaxing without disrupting influences: this can be done particularly well in rooms with delicate, unobtrusive colours and fabrics. Haptic stimuli, which address and simultaneously soothe the senses, are as important in this respect as the colourfulness described above. **Landgasthof Bären** in Schura is a good example of this, especially the "Natur Pur" rooms, which deliver the "pure nature" promised by their name in a relaxing, down-to-earth and simultaneously modern way (interior design: hotelident Michaela Voß).

以柔和、不显眼的色彩和面料设计的房间尤能使人觉得放松，不易受干扰。可舒缓情绪的触觉刺激在这方面同样重要。巴雷兰德斯加特霍夫酒店（Landgasthof Bären）的房间就是一个很好的例子，尤其是它的"纯天然"房，如名字一样，它通过放松、接地气而不失现代的设计手法提供了纯天然的享受（室内设计：hotelident Michaela Voß）。

A compelling design concept works whenever it is implemented on all levels, and in wellness areas, this also includes the corridors. When the guest walks along the pictured corridor in the **Baltic Sea Grand Spa of the Steigenberger Hotel in Heringsdorf,** he is enveloped by walls illuminated with warm light, which lead him along an elegantly curved passageway. The language of shapes in the entire spa area is plain, the colours are frequently reduced to the natural colour of the materials used or to cream hues, so that a warm and relaxing atmosphere is generated. Colour accentuations – which all correspond to the colour concept – can be controlled by the guests themselves depending on their taste and mood, whilst having a bath in an exclusive massage pool for two in a spa suite: from invigorating green to lilac to blue light, everything is possible (interior design: JOI-Design).

如若酒店有引人入胜的设计理念，不管设计到哪种程度，健身区都会有走廊。当客人沿着黑林格斯多尔夫斯泰根博格温泉大酒店（Steigenberger Grandhotel & Spa Heringsdorf）挂满画的走廊散步时，墙壁被温暖的光线照亮，客人可沿着曲折而雅致的廊道前行。整个SPA区域的造型非常平常，色彩也简化为所用材料的自然色彩或奶油色，从而营造温暖和轻松的氛围。与其色彩理念相呼应，客人可以自己的情绪和品位控制重点色彩；同时，水疗套房中设有独家按摩池，想调成活泼的绿色、丁香紫色还是淡蓝色？一切皆有可能（室内设计：JOI-Design）。

However, public wellness areas are not the only places promising relaxation: in the **Mövenpick Hotel Munich Airport**, guests have their own private little spa area in their bathroom, accented with a crunchy apple green. The nature motif – symbolised by fresh green – allows business guests at the airport to relax and recharge their batteries for the next day (interior design: JOI-Design).

Adex Silver Award 2009

然而，公共健身区域并不是唯一可放松的地方：**慕尼黑机场莫凡彼酒店**（Mövenpick Hotel Munich Airport）客房的浴室中设有小型私人SPA区域，主要使用了新鲜的苹果绿色。清新的绿色象征了机场酒店的自然主题，使得商务旅客可以放松、为第二天充电（室内设计：JOI-Design）。

曾获2009年Adex大奖银奖。

Guests can also relax in an exclusive atmosphere characterised by a slightly bolder colour concept in the wellness area or the spa suite of **Hotel Ritter** in Durbach. Everything is designed in shades ranging from cream to sand to grey, and reflects the local natural environment. Bright pink and green as accent colours add visual highlights. In the relaxation areas and suites, fireplaces and artworks with references to the region reinforce the comfortable atmosphere. The design concept for the hotel with its adjoining vineyards repeatedly takes up the theme of "wine" with sophisticated details, and so one comes across Riesling green next to various nuances of Pinot Noir red running through the hotel like a unifying theme (interior design: JOI-Design).

里特尔杜尔巴赫酒店（Hotel Ritter Durbach）的客人还可以在健身区或水疗套房放松，它们有着独特的氛围和稍大胆的色彩概念。一切都被设计成奶油色调、沙色或灰色，深浅不等，反映了当地的自然环境。明亮的粉红色和绿色作为强调的色彩增添视觉亮点。休闲区和套房中，与当地民俗文化相关的壁炉和艺术品强化舒适的氛围。酒店的设计理念与相邻的葡萄园有着密切的联系，"酒"的主题和复杂的相关技术细节多次出现，雷司令葡萄酒的绿色和各种比诺葡萄的红色作为统一的色彩主题贯穿整个酒店（室内设计：JOI-Design）。

A young, fresh concept: an organic language of forms and untreated surfaces together with the colour canon characterise the design idea for the **Park Inn Nürnberg.** Owing to the dominant and very fresh green, the depicted bar at the Park Inn Nürnberg inevitably feels vitalising – even without naming the colour world – and conveys a sense of liveliness to the guests (interior design: JOI-Design).

纽伦堡公园酒店（Park Inn Nürnberg）的设计综合了有机的形式、未经处理的表面和标准的色彩。这是一种很新的理念。由于新鲜的绿色占主导地位，酒店的酒吧让人觉得活力四射，即使它的色彩世界未作进一步处理，也为客人传递了活力（室内设计：JOI-Design）。

Dolce Munich	authentic / 返璞归真	III	174 + 175, 176 + 177		JOI-Design
Privathotel Hannover	authentic / 返璞归真	III	178 + 179		JOI-Design
Les Jardins d'Alysea	restrained / 婉约矜持	III	190 + 191		JOI-Design
art'otel Köln	restrained / 婉约矜持	III	192	bottom left / 左下	Christoph Kraneburg, Köln
Hotel Otterbach	restrained / 婉约矜持	III	192	top left / 左上	Christoph Kraneburg, Köln
Hotel Otterbach	restrained / 婉约矜持	III	193	right / 右	Christoph Kraneburg, Köln

The **Dolce Munich Hotel** in Unterschleißheim is the epitome of "authenticity": a reduced and exclusive material mix consisting of native woods, bronzed metal, polished fair-faced concrete, leather, loden, and hides establishes credible connections to the region. The business hotel stands for a traditional Bavarian lifestyle, combined with contemporary details. Contemporary shapes mixed with untreated natural materials express exclusiveness, style and individuality. In the lounge, the natural colour spectrum is interrupted by a single individual accents, like a raspberry-coloured table. In the hotel's "Redox" gourmet restaurant, the ambience is characterised by muted colours and subtle anecdotes from the Bavarian surroundings, like chandeliers made from deer antlers, combined with oak with a light or dark finish. Additionally, the interplay of direct and indirect light creates exciting bright-dark contrasts (interior design: JOI-Design).

Finalist: European Hotel Design Awards 2010

翁特尔斯希莱斯海姆的**慕尼黑杜尚酒店**（Dolce Munich Hotel）是"地道原味"风格的典型，它混合使用了当地独特的、加工较少的材质，如原生木材、古铜色金属、浅色抛光混凝土、皮革、洛登缩绒防水厚呢和兽皮。该商务酒店将传统的巴伐利亚生活方式和现代的细节相结合。未经处理的天然材料和现代的造型结合显得尊贵、富有格调和个性。休息室内，桌子等的树莓色作为重点色彩，点缀了整体的自然色彩。杜尚酒店内的"Redox"美食餐厅有着鹿角制成的吊灯、橡木与浅色或深色面漆，通过柔和的色彩和有趣的巴伐利亚风格内饰营造了氛围。此外，直接和间接光的相互作用营造出令人兴奋的明暗对比（室内设计：JOI-Design）。

入围：2010年欧洲酒店设计奖。

The colour concept of the rooms in **a privately-run hotel in Hanover** also features the world of "authentic" colour in a modern interpretation: tables and wardrobes in various shades of brown, yellow, black, and light grey are combined in the carpet pattern. A design concept between tradition and current zeitgeist: the noblesse of past times was closely connected to the horse, the Hanoverian – the gentleman rode a horse, he did not walk. Design details represent references to bridles and leather in a contemporary context (interior design: JOI-Design).

汉诺威私人酒店（Privathotel Hannover）的客房是对"地道原味"风格色彩的现代诠释，褐色、黄色、黑色、浅灰色等各种色调的桌子和衣柜上都使用了地毯上的图案。该酒店的设计理念综合了传统和现代的时代精神。可以说，汉诺威过去的贵族生活和马匹紧密相连，过去的汉诺威绅士往往骑马而不是走路，因而酒店在现代环境中采用了一些笼头和皮革的设计细节（室内设计：JOI-Design）。

Les Jardins d'Alysea: the colour world of this pleasantly "restrained" design shows itself to elegant effect. Subtle shades of grey, beige and lilac in combination with materials with a very pleasant surface feel and the correct light unfold an impressively restrained and simultaneously homely effect in the spacious rooms. Subtle floral motifs on walls, cushion covers and upholstery fabrics further loosen up the colour scheme and give it a masterly touch of liveliness. Roses were once cultivated here. The interior design lovingly plays with the theme, and many a detail reveals itself to the beholder only at second glance (interior design: JOI-Design).

阿莉莎花园酒店（Les Jardins d'Alysea）："内敛"而怡人的色彩散发出优雅的气息。淡淡的灰色、米色和丁香紫色调和表面质感颇佳的材质相结合，还有怡人的光线，在宽敞的房间里展示出低调而亲切、令人印象深刻的效果。这里曾经种植玫瑰花。墙壁、垫子和装饰面料上精美的花卉图案调和了配色方案，巧妙地增添了些许生动活泼的气息。室内设计精心呼应主题，其中的许多细节旁观者一两眼便能参透一二（室内设计：JOI-Design）。

The interiors of the **art'otel Köln** and **Hotel Otterbach** likewise show how stylish "restraint" can be. Extensive carpet designs in various shades of grey appear pleasantly unobtrusive and attract enough attention so that they blend in with the overall ambience. Skilful colour accentuations with delicate yellow and lilac hues give the ambience warmth (interior design art'otel Köln: NALBACH + NALBACH Gesellschaft von Architekten mbH; interior design Hotel Otterbach: Ziefle Koch GmbH)

科隆艺术酒店（art'otel Köln）和**奥特巴赫酒店**（Hotel Otterbach）的室内同样展示了如何时尚而"内敛"。各种灰色色调、不断延伸的地毯令人惊喜，虽不张扬但能吸引足够的注意力，且融入整体的氛围。娴熟地设计重点色彩，与淡淡的黄色和淡紫色色调结合，营造出温暖的氛围（科隆艺术酒店室内设计：NALBACH + NALBACH Gesellschaft von Architekten mbH；奥特巴赫酒店室内设计：Ziefle Koch GmbH）。

Cowley Manor	family-friendly / 居家温情	III	204 + 205		ege®
Radisson BLU Hotel Amsterdam Airport	puristic / 纯粹至美	III	217		Arne Jennard
Radisson BLU Hotel Amsterdam Airport	puristic / 纯粹至美	III	218	top left / 左上	Arne Jennard
Hilton Munich Park	puristic / 纯粹至美	III	218	bottom / 下	JOI-Design
Hilton Munich Park	puristic / 纯粹至美	III	219	right / 右	JOI-Design
Adlon Hotel Berlin (Musterzimmer)	historic / 历史沉淀	III	232	bottom / 下	ABB Asea Brown Boveri Ltd. Busch-Jaeger Elektro GmbH
Grand Hotel Wien	historic / 历史沉淀	III	234 + 235		JOI-Design

Family-friendly is what is fun for the whole family: generous rooms, a playful and colourful design – particularly for the younger guests – as well as warm and touchable, relatively hard-wearing materials. The **Cowley Manor** hotel is certainly not conceived as a boundless playground for the youngest of guests; nevertheless, it combines the highest of standards and exclusivity with a family-friendly environment due to the aforementioned colourfulness and generous room design within an otherwise reduced, modern interior (interior design: De Matos Ryan Architects).

所谓有利于家庭生活的，即可为整个家庭带来乐趣，如宽敞的房间，好玩的、丰富多彩的设计（尤其是为年轻客人准备的），温暖、有质感，以及相对耐磨的材料。虽然考利园酒店（Cowley Manor）肯定不会被小客人们视作无边的游乐场，但它仍然可满足最高标准，尽显尊贵；正如我们前面所提及的，丰富多彩而宽敞的客房设计和现代简约的室内营造出家一般的氛围（室内设计：De Matos Ryan Architects）。

Puristic interior design cannot be equated with a colourless approach. Especially the reduced and straight-lined requires skilful coloured accentuations, as can be found in the design of the rooms in the **Radisson BLU Hotel Amsterdam Airport**: austere forms, further emphasised with plenty of black, grey and definitely also white, are broken up by plain but radiant yellow decorative cushions and an equally yellow desk chair. Owing to its balance between grey and warm wood, the room exudes the calmness the guests at the airport long for (concept & design: Creneau International).

简洁的室内设计并不等同于无色，简约和线形的设计更需要重点色彩的熟练运用。例如，阿姆斯特丹机场雷迪森酒店（Radisson BLU Hotel Amsterdam Airport）的客房使用了大量的黑色、灰色，当然，还有白色，突出了简朴的造型，样式简单而富有活力的黄色装饰靠垫和同样的黄色桌椅点缀其中。由于灰色和温暖木质之间的平衡，房间有着机场旅客所需的冷静气质（概念与设计：Creneau International）。

The example of the room and suite designs in the **Hilton Munich Park** clearly illustrates that puristic does not necessarily mean straight: a desk with lathed legs and a black desktop also looks puristic in the appropriate environment. In the design concept for the hotel located directly at the English Garden, the overall impression is shaped by sophisticated Bavarian references and the nearby natural landscape, with its floral formal elements (interior design: JOI-Design).

Finalist: European Hospitality Award 2012

慕尼黑公园希尔顿酒店（Hilton Munich Park）的客房和套房则向我们示范了简洁的设计并不意味着色彩单一或直线形，曲线造型的桌腿和黑色的桌面看起来依然简洁。由于酒店位于英式花园内，它的设计理念是将提炼的巴伐利亚风格和附近自然景观中的花卉形状元素相结合（室内设计：JOI-Design）。
入围：2012年欧洲酒店大奖。

Historic design, on the other hand, is characterised by muted and mostly rich brown, red, beige, and also golden shades. It draws on modern or also stylised interpretations of furniture and patterns from bygone days. The **showroom in the Adlon Hotel in Berlin** represents the world of "historic" colour, which is staged by means of skilfully positioned light. The flair of bygone days and the hotel's famous history are revived and paid tribute to in the design concept (interior design: LIVING DESIGN).

另一方面，它柔和、丰富而深浅不一的棕色、红色、米色的设计富有历史感。设计借鉴了往日的家具和图案，但注入了现代的诠释和格调。柏林阿德隆酒店的陈列室（Adlon Hotel Berlin (Musterzimmer)巧妙地借助定位灯，展示了所谓"历史悠久"的色彩。设计理念重燃并歌颂了酒店往日的风情和鼎鼎有名的历史（室内设计：LIVING DESIGN）。

In the room of the **Grand Hotel Wien** shown here, the colour emphasis is on brighter shades, the red, beige and golden yellow ones. The patterns of the fabrics, carpets and wall coverings leave no doubt about which design theme has been selected here. Details such as dark wooden chairs with a floral jacquard fabric, which is separated from the baroque ornament of the chair frame by golden metal rivets, further emphasise this. Through the design concept for the rooms and suites in the Grand Hotel, which is located close to the Vienna State Opera House, the history of Princess Elisabeth lives on, and guests are put in the right mood for the city with countless historical sites (interior design: JOI-Design).

维也纳大酒店（Grand Hotel Wien）的客房重在红色、米色和金黄色等亮色调。织物、地毯和墙面的图案都清楚明了地展示了设计主题。深色木椅子上的花卉提花面料，椅子的金色金属铆钉框架和巴洛克风格的饰物，进一步强调了这一点。大酒店和史上伊丽莎白王后曾居住的维也纳国家歌剧院相邻，客房和套房的设计使得客人们不禁沉浸在这座历史遗迹无数的城市里（室内设计：JOI-Design）。

Restaurant	romantic / 浪漫多情	IV	249		Panaz
Le Méridien Stuttgart	poetic / 朦胧诗意	IV	262	top left; bottom / 左上；下	Le Méridien Stuttgart
Art déco (Zimmer)	poetic / 朦胧诗意	IV	263	top centre; right / 中上；右	JOI-Design
Hotel Gendarm nouveau	imaginative / 想入非非	IV	274 + 275		JOI-Design
Hotel Ritter Durbach	inspiring / 心驰神往	IV	286	bottom left / 左下	Hotel Ritter Durbach
Hotel Ritter Durbach	inspiring / 心驰神往	IV	286	top left; centre /左上；中	JOI-Design
Hotel Ritter Durbach	inspiring / 心驰神往	IV	287		Hotel Ritter Durbach

The world of "romantic" colour is characterised by gentle nuances, as is illustrated by the **view of the restaurant**. Even though playful shapes and soft materials belong to such an ambience, one should one should refrain from including too many frills and dainty decoration in all colours of the spectrum. A harmonious whole as shown in this example is generated by calm fabrics, masterly accentuations – like the flower-like hanging lamps, which additionally give off warm light – and artworks on the walls complementing the colour concept.

"浪漫多情"，如249页图中的**餐厅**（Restaurant）所示，特点在于色彩细微而温柔的差别。即便这样的氛围容许有趣的形状和柔软的材料，还是应避免包括太多色彩多样的褶边和纤巧的装饰。如图，在这个和谐的整体中，稳重感十足的面料衬托了餐厅的重点色彩所在——散发暖暖光线的花形吊灯和墙上的艺术品。

Poetry envelopes the longing for lightness and for beautiful things. The **Le Méridien Stuttgart** fulfils this desire with rooms characterised by a very bright and warm design: wallpaper with modern, slightly abstracted flowers in subtly hues, a subtle patterned carpet, clear and yet soft furniture contours. The view from the balcony of the Schlossgarten became the inspiration for design ideas: modern interpretations of motifs from the palace bring historic patterns on wallpapers and fabrics into the modern era (interior design: JOI-Design).

斯图加特艾美酒店（Le Méridien Stuttgart）的诗意含着对明朗和美丽事物的向往。酒店以非常明亮和温暖的设计满足人们的这一愿望。墙纸上有着现代而稍显抽象的花纹，色调柔和，地毯上的花纹也不显眼，家具轮廓清晰而又柔和。从城堡花园的阳台上欣赏美景成为设计思路的灵感来源，设计对宫廷主题的壁纸和织物图案进行了现代的诠释，带来了现代气息（室内设计：JOI-Design）。

The **Art Déco room** shows a design from this colour world, which reveals an unexpected interpretation with noticeably more grey nuances. Despite the defining silver grey of the wallpaper, but adorned with large numbers of flowers, offset in bright nuances, a fine air of poetry is breathed into the room. Much influenced by the lily, the design concept stands for the consistent implementation of a motif right down to the last detail – a hotel that promotes recollection and relaxation (interior design: JOI-Design).

装饰艺术室（Art Déco (Zimmer)）展示了诗意风格的色彩设计，令人意外的是它使用了更微妙的灰色。尽管壁纸定义为银灰色，但以大量鲜花装饰，色调略微向亮色偏移，房间中洋溢着诗意。百合花多次出现，这种设计理念始终贯穿如一，直至最后的细节。这是一家可以唤起回忆、可以放松的酒店（室内设计：JOI-Design）。

At the **Hotel Gendarm nouveau** in Berlin, guests encounter an inspiring, imaginatively designed ambience. Muted jacquard fabrics with modern floral patterns offset in pink, decorative cushions in blackberry and olive green, beige-coloured walls, some with contemporary takes on pink-tinted depictions of silhouette portraits which in turn are framed by opulent ornaments. The interior as reminiscence of the neighbouring Gendarmenmarkt: contemporary colours and surfaces reinterpret contemporary history and take it to the "here and now" (interior design: JOI-Design).

在柏林的新皇家卫士酒店（Hotel Gendarm nouveau），客人可感受到鼓舞人心的、富有想象力的设计氛围。柔和的提花面料、粉红色而不失现代的花卉图案、黑莓色装饰靠垫、墙壁上橄榄绿色与米色相间，现代的粉红色花纹和华丽的内饰相得益彰。酒店室内让人想起了相邻的御林广场，设计通过现代化的色彩和外观重新诠释了历史（室内设计：JOI-Design）。

The gourmet restaurant at the **Hotel Ritter** in Durbach has an inspiring effect on the guests, while the surrounding area was used as a source of inspiration for the design. Wallpaper made specifically for this project extends over a complete wall; its design reacts to the theme of "wine", which plays an important role due to the hotel's location. The theme of wine flows through the entire building as an inspiration for the shapes and colours, and provides the connecting design element (interior design: JOI-Design).

里特尔杜尔巴赫酒店（Hotel Ritter Durbach）美食餐厅的设计灵感来自于周边，富于启示性。壁纸专门为这个项目订制，覆盖了整面墙；由于酒店的位置原因，"葡萄酒"的主题贯穿始终，设计也与其呼应。葡萄酒主题流经整个建筑，成为形状和颜色的灵感来源，并提供了相关的设计元素（室内设计：JOI-Design）。

Steigenberger Grandhotel & Spa Heringsdorf	classic / 经典传承	IV	298 + 299, 300 + 301		JOI-Design
Sheraton Hannover Pelikan Hotel	classic / 经典传承	IV	302 + 303		JOI-Design
Fährhaus Munkmarsch	snug / 舒适温暖	IV	314	centre left / 左中	Ydo Sol
Fährhaus Munkmarsch	snug / 舒适温暖	IV	314	top left; centre / 左上；中	JOI-Design
Fährhaus Munkmarsch	snug / 舒适温暖	IV	315		Ydo Sol
Europäischer Hof	snug / 舒适温暖	IV	316		JOI-Design
Hotel Ritter Durbach	snug / 舒适温暖	IV	317	bottom left / 左下	Hotel Ritter Durbach
Hotel Ritter Durbach	snug / 舒适温暖	IV	317	right / 右	JOI-Design

Owing to their elegant plainness, classic shapes and colours exude calm and serenity and invite one to leave one's cares behind. In the depicted apartment of the **Steigenberger Grandhotel & Spa Heringsdorf**, guests are received by exactly this atmosphere. Colours like red and grey as well as modern furniture made from walnut-titanium stained wood or lights with bright, rectangular fabric lampshades and chrome-plated frames give the apartment a classic, international, inviting style. This concept was also implemented in the public areas: the reception area was designed in bright colours, and warm light fills the room. The reception desk is complemented with a table, where all check-in formalities can be completed while comfortably seated. A fireplace is located further back, which is integrated as a central element into a stone wall cladding reminiscent of amber and the beach. The idea behind the design concept was to create a warm, familiar reception with repeated references to the history of the location (interior design: JOI-Design).

黑林格斯多尔夫斯泰根博格温泉大酒店（Steigenberger Grandhotel & Spa Heringsdorf）平实、典雅的造型和色彩使人觉得平静与安宁，烦恼都可以抛之脑后。如图所示，公寓式客房带给客人的正是这种氛围。红色和灰色、现代的有色核桃木和钛合金家具、明亮的灯光、长方形布灯罩与镀铬框使得公寓式客房看起来友好、经典而具有国际范儿。公共场所也以此理念设计：接待区采用明亮的色调，温暖的灯光充满了房间。前台处另有一张桌子，客人办理入住时可舒舒服服地坐在上面。壁炉在后面较远处，它和石头墙面相融合，变成了墙面的中心元素，让人联想起琥珀和海滩。此设计背后的理念是营造温暖的接待过程，并使得人们熟悉酒店的历史（室内设计：JOI-Design）。

A modern variation of the classic is presented by the **Sheraton Hannover Pelikan Hotel**: plenty of brown, black and grey with accentuating red and yellow in the carpet, the wall coverings and the pieces of furniture like the headboard of the bed. The indirectly illuminated artwork on the headboard is an eyecatcher. As a homage to "Pelikan" it shows a fountain pen and handwritten notes. The overall design is much influenced by the traditional writing utensil, which was once produced here. That is why the hanging lamps are reminiscent of inkpots and the edge of the side table is reminiscent of the typical jagged inkpot lid (interior design: JOI-Design).

汉诺威喜来登佩利坎酒店（Sheraton Hannover Pelikan Hotel）为经典注入现代的诠释，呈现在人们面前的是大量棕色、黑色和灰色，与地毯、墙面覆盖物和床头家具上突出的红色和黄色相配。床头隐隐照亮的艺术作品非常吸引人，为了向"佩利坎"致敬，作品上呈现了一支钢笔和手写的笔记。由于这里曾是传统书写工具——钢笔的产地，整体设计受传统的书写工具的影响。这就是为什么吊灯让人想起墨水瓶，而靠墙桌子边缘使人联想到典型的锯齿墨水瓶的盖子（室内设计：JOI-Design）。

Natural and warm shades in particular are "snug" – the vibrant ones are used only to add accentuations. At the **Fährhaus Munkmarsch Hotel**, this colour concept has been brilliantly implemented: the common theme of all the rooms is the nature of the island of Sylt, and the guests can feel it as soon as they open the door. The warm beige and brown colours of the choice materials are reminiscent of the beach and give the rooms a quaint atmosphere. Like sand with its fine structure, the surfaces of the pieces of furniture, which vary slightly in their colour nuances and are partly made of brushed wood or textile covers, have a very pleasant structure to the touch. Special details are, amongst others, the small leather-covered reading lamps at the head of the bed or the modern interpretation of a rocking chair with lambskin (interior design: JOI-Design).

"温暖舒适"的色调自然、温暖而有活力，可以突出温暖舒适的氛围。蒙克马施法瑞汉斯酒店（Fährhaus Munkmarsch Hotel）出色地实施了这种色彩理念，所有客房的共同主题是叙尔特岛的大自然，客人一打开门，便能感受到它的存在。选料温暖的米色和棕色让人联想到沙滩，并为房间营造古朴的氛围。家具颜色略有不同，部分采用拉丝木材或纺织品覆盖，表面有着令人愉快的触感，摸起来有些沙子的感觉。其中有些特别的细节，如床头的阅读灯以小皮革覆盖，摇椅覆以羔羊皮显得更现代（室内设计：JOI-Design）。

The room at **Europäischer Hof** is also consistent with the snug colour scheme, whereby fewer beige hues and more light brown and cognac shades were applied, which immerse the rooms in a warm light when the sun is shining. The interior does without excessive or opulent decoration and also without gaudy colours. The design for the city hotel in the centre of Hamburg gives priority to a conservative, calm approach: here, the guests can relax from the hustle and bustle outside the main entrance and regain their strength for the next day (interior design: JOI-Design).

温暖舒适的色彩方案在欧罗巴斯赫尔霍夫酒店（Europäischer Hof）的客房中贯穿始终。它运用了更少的米色色调、更多的浅棕色和白兰地色调，这使得房间在阳光明媚时会沉浸在温暖的光线中。内饰确实没有过多华丽的装饰，也没有花哨的色彩。由于这家酒店位于汉堡的市中心，设计师优先考虑保守、宁静的设计手法，在这里，客人可以远离店外的喧嚣，放松恢复体力，迎接第二天的到来（室内设计：JOI-Design）。

The design concept for the rooms at **Hotel Ritter Durbach** illustrates how ingenious effects can be achieved with just one subtle coloured detail: the floral motif on the chair adds a radiant emphasis and simultaneously offers a haptic experience. The bathroom picks up the colour canon in its warm tiles (interior design: JOI-Design).

里特尔杜尔巴赫酒店（Hotel Ritter Durbach）的设计理念说明了如何通过微妙的色彩细节获得巧妙的效果：椅子上的精美花卉图案带来了色彩重点，同时提供了触觉体验。浴室暖色调的瓷砖使浴室不再沉闷（室内设计：JOI-Design）。

Messe GAST (Voglauer)	nonconformist / 不入主流	V	330 + 331		Voglauer hotel concept
Le Clervaux Boutique & Design Hotel	bohemian / 放荡不羁	V	342	top left / 左上	Le Clervaux Boutique & Design Hotel
Le Clervaux Boutique & Design Hotel	bohemian / 放荡不羁	V	342 + 343		JOI-Design
Hilton Vienna Danube	chic / 潇洒别致	V	354 + 355		JOI-Design
Temple Restaurant Beijing	chic / 潇洒别致	V	356		ege® GmbH
Le Méridien Brüssel	chic / 潇洒别致	V	358 + 359		Lutz Voderwühlbecke

The room with a nonconformist design **(Messe GAST)** surprises guest with new shapes and experimental combinations, which are not held hostage by tradition! Styles are mixed as freely as the materials and colours from the entire range of terms. This room is intended to distract its user from everyday routines, offer a change, galvanise him, and inspire him – a polarising concept (design & concept: Voglauer hotel concept).

Messe GAST房间（Messe GAST (Voglauer)）将新颖的形状和实验造型相结合，充分表明它的设计不墨守成规，不受传统限制，可为客人带来惊喜！从材料到颜色，整个范围内风格都自由混合在一起。这个房间是为了吸引用户从日常惯例中走出来，它提供了变化和两极分化的概念，激励用户，启发用户（设计与理念：Voglauer hotel concept）。

The "fine dining area" of the restaurant at the **Le Clervaux Boutique & Design Hotel** is distinguished by the tasteful, modern interpretation of baroque patterns, shapes and colours, which draw one the palace opposite the hotel for its design concept. Red, beige and light grey in the carpet design – it is based on a pattern of historic tiles which lined the hotel corridors in the past – and the dark anthracite hue of the high-quality curtains give the room a glamorous background. The upholstered chairs covered in vibrant red and noble black with an opulent pattern are perfectly integrated into the ambience, which unfolds its full effect especially in the evening because of its ample indirect lighting (interior design: JOI-Design).

Boutique Media: Boutique Design Award – Most surprising visual element 2012. Heinze Verlag: Heinze Publikums Preis 2013

勒克莱沃精品设计酒店（Le Clervaux Boutique & Design Hotel）"精致的餐饮区"的设计理念受到酒店对面宫殿的启发，但对巴洛克风格的图案、形状和色彩进行了现代诠释，显得格外雅致。地毯上的红色、米色和浅灰色的图案基于过去酒店走廊的瓷砖图案而设计，无烟煤暗色调的优质窗帘为房间带来迷人的背景。凭借鲜艳的红色和高贵的黑色，以及华丽的图案，软垫椅子完美地融入了氛围，尤其是在晚上，由于充足的间接照明，它们可以展现出全部魅力（室内设计：JOI-Design）。

曾获2012年精品酒店设计大奖 / 最令人惊艳的视觉元素奖；并刊载于海因策出版社（Heinze Verlag）的《Heinze Publikums Preis 2013》。

In the guest rooms of the **Hilton Vienna Danube,** the large headboard's supple leather cover with undulating stitching is combined with light grey and various red and lilac-coloured nuances on cushions and the carpet. The charm of the colour composition is emphasised with straight lines, which comply with the zeitgeist and the term "chic". The design concept plays with the movement of waves of the Danube flowing past and refers to it in many small details (interior design: JOI-Design).

维也纳希尔顿多瑙河酒店（Hilton Vienna Danube）客房中床头板的皮革有着波浪形的色彩拼接，它结合了坐垫和地毯上的浅灰色、各种红色和丁香紫色。直线形状更能凸显色彩组合的魅力，既符合时代精神，也显得"时尚"。客房设计理念围绕流经酒店的多瑙河，许多小细节也与多瑙河的涟漪息息相关（室内设计：JOI-Design）。

In the **Temple Restaurant Beijing**, plain but rounded shapes, like those that characterise the various seats and chairs, are enveloped with a harmonious colour canon of grey, red and rust-brown upholstery fabrics. Superfluous or even opulent decoration was deliberately omitted. The concept banks on "pure elegance" and thus allows the pieces of art in the room to unfold their special effect on the guest (interior design: HASSELL).

北京古寺餐厅（Temple Restaurant Beijing）的各种座椅、椅子都采用了朴实而圆润的形状，座套面料则采用了协调的灰色、红色和锈褐色。多余甚至是奢华的装饰被有意省略。设计理念为"纯粹的优雅"，从而使得客人可发现房间中艺术品的独特魅力（室内设计：HASSELL）。

The room scenario from the **Le Méridien Brussels** reveals a different side of the "chic" interior environment: surrounded by a calm beige-brown context, orange-coloured, elegant little tables and various red shelves on the walls with all sorts of collected objects, like a globe, books and small pots, allow a modern living room atmosphere to develop. Despite this generous decoration, the room does not appear less harmonious, but well co-ordinated (interior design: Marc Humbert – Brussels).

布鲁塞尔艾美大酒店（Le Méridien Brussel）的室内场景则揭示了"时尚雅致"的另一面：它为冷静的米色、棕色所包围，优雅的橙色的小桌子和墙壁上摆满地球仪、书籍和小花盆等各种收藏物的红色支架，营造现代客厅的氛围。室内显得非常宽敞大方，且非常和谐（室内设计：Marc Humbert – Brussel）。

Mercure Hotel & Residenz Frankfurt	richly accentedd / 浓墨重彩	V	370 + 371	JOI-Design
Organic Trace	futuristic / 未来主义	V	382	Tobias D. Kern
Organic Trace	futuristic / 未来主义	V	383	JOI-Design
Park Inn Krakau	futuristic / 未来主义	V	384 + 385	JOI-Design

The **Mercure Hotel & Residenz Frankfurt** takes harmonious and contrary positions in its design: the "coolness" of the metropolis is contrasted with a distinctive colour accentuation in orange, which is integrated into a grey-brown colour context, creating cosiness and warmth. The architecture has freed itself from the right angle, it is sweeping and dynamic, which is additionally emphasised by the warm accent colour (interior design: JOI-Design).

法兰克福美居酒店（Mercure Hotel & Residenz Frankfurt）的设计考虑了对比和协调：大都市感的"冷色调"和鲜明的橙色形成对比，在灰棕色的背景下，营造舒适和温馨的氛围。室内从正确的建筑角度中解脱出来，有着"横扫千军"般的动感，从强调使用的暖色调便可见一斑（室内设计：JOI-Design）。

The **Organic Trace** room concept combines the living areas of "business" and "leisure" in a futuristic way: the centrally positioned white organic sculpture accommodates the bathtub, from which the desk swings out like a wave and finally ebbs away to merge into the bed. The light blue-grey carpet and the petrol-coloured walls flow around the figure like a pebble in the riverbed – a room concept for the future (interior design: JOI-Design).

Organic Trace酒店（Organic Trace）的房间设计理念将 "商务"区和未来主义风格的"休闲"生活区融为一体：中央有着白色有机雕塑造型的浴缸，而桌子仿佛波浪一样摆动，最后退至床畔，与床合为一体。这是非常超前的房间设计理念，浅蓝色的地毯和汽油色调的墙壁围绕着雕塑，令人想起河床中的鹅卵石（室内设计：JOI-Design）。

In a similarly futuristic way, the reception area of the **Park Inn Krakow** receives the hotel guests with a spatial sculpture, which leads to the interior with a flowing gesture. Designed in impressive black and additionally framed with indirect ceiling lighting, the reception area is the core element of the Polish hotel. The design concept also reflects the strong architectural language in the interior: exterior and interior merge into one idea (interior design: JOI-Design together with Ovotz design Lab. and J. Mayer H.).

Boutique Design: Best futuristic design scheme 2009

克拉科夫公园酒店（Park Inn Krakow）的接待区同样呈现出未来主义风格，迎接客人的空间雕塑带给室内流动的姿态。接待区使用了令人印象深刻的黑色，在间接天花照明的勾勒下，成为这家波兰酒店的视觉中心。室内强有力的建筑语言也可以反映它的设计理念：外观和内饰融为一体（室内设计：JOI-Design together with Ovotz design Lab. and J. Mayer H.）。
曾获2009年精品酒店设计大奖最佳未来主义风格设计方案

Lindner Park-Hotel Hagenbeck	Oriental / 东方情调	VI	411	JOI-Design
Le Méridien Parkhotel Frankfurt	Oriental / 东方情调	VI	412 + 413	JOI-Design
Lindner Park-Hotel Hagenbeck	African / 粗犷非洲	VI	414 + 415	JOI-Design
Steigenberger Grandhotel & Spa Heringsdorf	North European / 纯美北欧	VI	426 + 427	JOI-Design

At the **Lindner Park-Hotel Hagenbeck,** distant worlds are suddenly very close: oriental blazes of colours paired with Far Eastern hospitality are perceptible inside the rooms, which – equipped with plenty of authentic furnishings and decorative fabrics in pink, orange and golden brown shades – carry the guests off to faraway places. In the hotel at the Hamburg zoo, the animals' countries of origin become accessible: the design concept invites guests to go on a journey of discoveries. In the suites, a colourful selection of fabrics highlights the Far Eastern character (interior design: JOI-Design).

hotelforum: Hotelimmobilie des Jahres 2009

在哈根贝克林德纳公园酒店（Lindner Park-Hotel Hagenbeck）里，遥远的东方世界突然变得非常接近，大量纷繁的东方色调和远东的热情好客呼之欲出，房间内配备了大量地道的家具和粉红色、橙色和金黄色色调的装饰性面料，仿佛带领客人到达了遥远的地方。酒店与动物王国——汉堡动物园相邻，客人很方便参观动物园，酒店的设计理念也旨在为客人开启一段发现之旅。套房中精选的多彩面料凸显了远东格调（室内设计：JOI设计）。
2009年酒店论坛：酒店物业奖。

In the bar of the **Le Méridien Parkhotel Frankfurt**, guests can enjoy not only the exotic ambience but, of course, also the corresponding drinks. Warm orange, brown and violet shades of the fabrics and the colourful, richly ornamented carpet designs perfectly complement the mosaics on the walls and tables, which capture the mat light of the ornate oriental hanging lamps. For the nightcap after a strenuous conference day, the design concept of this bar allows one to immerse oneself in a different world: warm and shining light reflections envelop the room and create highlights (interior design: JOI-Design).

Adex Platinum Award 2009

客人在法兰克福艾美公园酒店（Le Méridien Parkhotel Frankfurt）的酒吧不但可以感受异国风情，还可以品尝异国的饮品。温暖的橙色、棕色和紫罗兰色调，以及色彩、装饰都很丰富的地毯设计与墙壁和桌子的马赛克完美搭配，它们在捕捉着东方风格挂灯的灯光。这个酒吧的设计理念是使得温暖闪耀的光线笼罩着房间，成为设计亮点。经过煎熬的一整天会议，在酒吧里小酌一杯，人们便会沉浸在完全不同的世界（室内设计：JOI-Design）。
曾获2009年Adex大奖白金奖。

The trip to Africa continues: the theme rooms in the **Lindner Park-Hotel Hagenbeck** surprise guests with authentic details, which capture the atmosphere of the breathtaking "dark continent". A hand-woven, dark red decorative scarf with elephants, antelopes and lions is spread out across the bed in the Africa room. The headboard is finished with bright ostrich leather, and next to it, slender, manually lathed bedposts made from maple soar upwards. In the bathroom, lion heads as towel holders and a washbasin in a drum clad with dark coconut wood await the guests. The design concept for the guestrooms refers to the special location of the hotel directly next to the Hamburg zoo: the rooms prepare, the guests for Africa's history and a day amongst the animals (interior design: JOI-Design).

hotelforum: Hotelimmobilie des Jahres 2009

非洲之旅还在继续：哈根贝克林德纳公园酒店（Lindner Park-Hotel Hagenbeck）主题客房中真实的细节令客人惊讶，它还原了 "神秘大陆"令人惊叹的氛围。非洲风客房床边有着手工编织的暗红色的、以大象、羚羊和狮子为图案的装饰围巾。床头板表面采用亮鸵鸟皮革，在它旁边，枫木制成的细长床柱可以手动上升。浴室里的毛巾架、暗椰木洗脸盆上都有狮头图案。客房的设计理念体现了酒店与汉堡动物园毗邻的特殊位置，一天都在动物群中度过的客人可以事先熟悉非洲的历史（室内设计：JOI-Design）。
2009年酒店论坛：酒店物业奖。

Once again the **Steigenberger Grandhotel & Spa Heringsdorf**: As guests walk to the maritime-style rooms, the corridors with blue-and-white walls and ornaments resembling mooring ropes in the brown herringbone patterned carpet set the right mood for the room ambience. Large cushions, leaned against white wooden wainscotting, generate a warm, cosy atmosphere, complemented with white, dark blue and cream furniture as well as wallpaper with broad stripes in muted shades of blue. North European interior design could hardly be more inviting. With this interior-design concept, the new building on a historic site draws a line from the present to the past: the guest is in a pleasant way taken back to the era of former imperial seaside resort culture and is able to wind down from the pace of everyday life in a wonderful way (interior design: JOI-Design).

再来说一下黑林格斯多尔夫斯泰根博格温泉大酒店（Steigenberger Grandhotel & Spa Heringsdorf），海洋风的客房有着蓝白墙壁的走廊，和地毯上的人字形图案、类似缆绳的饰物一起营造了海上航行一般的室内氛围。白色护墙板上的大靠垫增添了温馨、舒适感，辅以白色、深蓝色和奶油色的家具以及壁纸，上面有着柔和的蓝色条纹。北欧风的室内设计则更加吸引人。在此室内设计理念的指导下，这个建于历史遗址上的新建筑向过去致敬，客人们可回顾帝国的海滨度假胜地文化，放慢日常生活的步伐（室内设计：JOI-Design）。

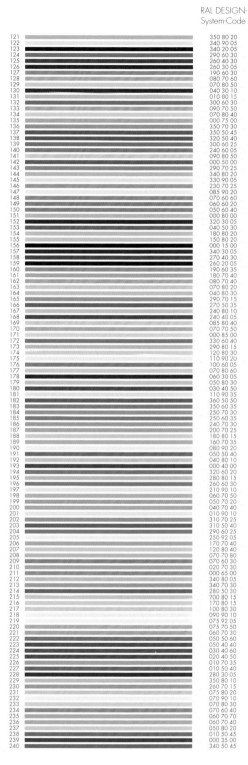

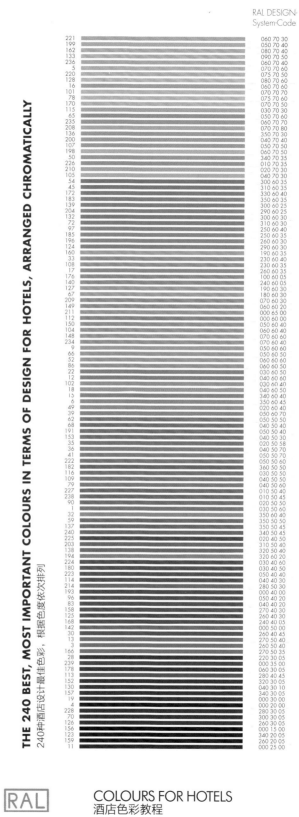

图书在版编目（CIP）数据

酒店色彩教程 /（德）维恩,（德）洛斯奇,（德）乔恩克著；深圳市艺力文化发展有限公司译. -- 南京：江苏凤凰科学技术出版社, 2016.1
 ISBN 978-7-5537-5316-4

Ⅰ. ①酒… Ⅱ. ①维… ②洛… ③乔… ④深… Ⅲ. ①饭店 – 建筑色彩 – 教材 Ⅳ. ①TU247.4

中国版本图书馆CIP数据核字(2015)第216909号

江苏省版权局著作权合同登记章字：10-2015-325号
COLOURS FOR HOTELS
© 2013
Text copyright Axel Venn, Corinna Kretschmar-Joehnk
Editor copyright Wolf D. Karl, RAL gemeinnützige GmbH
Layout, scientific preparation copyright Janina Venn-Rosky
Originally published in Germany in 2013 by Verlag Georg D.W. Callwey GmbH & Co. KG, Munich, Germany.
English translation rights originally arranged with Bianca Murphy, Murphy Translation Office, Hamburg, Germany.

酒店色彩教程

著　　　者	（德）阿克塞尔·维恩（Axel Venn）	
	（德）约阿尼纳·维恩·洛斯奇（Janina Venn-Rosky）	
	（德）科琳娜·克雷奇马·乔恩克（Corinna Kretschmar-Joehnk）	
译　　　者	深圳市艺力文化发展有限公司	
项 目 策 划	凤凰空间/彭娜	
责 任 编 辑	刘屹立	
特 约 编 辑	赵　萌	

出 版 发 行	凤凰出版传媒股份有限公司
	江苏凤凰科学技术出版社
出版社地址	南京市湖南路1号A楼，邮编：210009
出版社网址	http://www.pspress.cn
总　经　销	天津凤凰空间文化传媒有限公司
总经销网址	http://www.ifengspace.cn
经　　　销	全国新华书店
印　　　刷	深圳市新视线印务有限公司

开　　　本	889 mm×1 194 mm　1／12
印　　　张	37
字　　　数	222 000
版　　　次	2016年1月第1版
印　　　次	2016年1月第1次印刷

标 准 书 号	ISBN 978-7-5537-5316-4
定　　　价	628.00元（精）

图书如有印装质量问题，可随时向销售部调换（电话：022-87893668）。